Globe Health Pr

Family Living and Sex Education

GLOBE FEARON
EDUCATIONAL PUBLISHER
PARAMUS, NEW JERSEY

Paramount Publishing

Content Reviewers

Robert F. Adams is the Supervisor for Comprehensive Health Education in Miami, Florida.

Executive Editor: Joan Carrafiello
Project Editor: Laura Baselice
Production Manager: Penny Gibson
Senior Production Editor: Linda Greenberg
Art Director: Nancy Sharkey
Text, Cover, and Art Design: Armando Baéz
Manufacturing Supervisor: Della Smith
Photo Researcher: Jenifer Hixson
Marketing: Sandra Hutchison

Photo Credits: 1: Read D. Brugger, The Picture Cube; **3:** © David Booker; **4:** Ellis Herwig, The Picture Cube; **12:** Gloria Karlson, The Picture Cube; **14:** Steve Skjold; **15:** Steve Skjold; **19:** Steve Skjold; **24:** Spencer Grant, The Picture Cube; **26:** American Museum of Natural History; **39:** The Stock Shop; **41:** (left): Biosphere/Photo Researchers; (right): Nestle/Photo Researchers; **46:** Kindra Clineff, The Picture Cube; **50:** Alexander Tsiaras, Photo Researchers, Inc.; **54:** Milton Feinberg, The Picture Cube; **57:** Barbara Reis/Photo Researchers; **60:** Susan Van Etten, The Picture Cube; **61:** Scott Camazine, Photo Researchers, Inc.; **62:** (left) Teri Leigh Stratford, Photo Researchers, Inc.; (middle) Susan Van Etten, The Picture Cube; (right): William Thompson, The Picture Cube; **63:** Susan Van Etten, The Picture Cube; **64:** William Thompson, The Picture Cube; **65:** UPI/Bettmann; **73:** David Strickler, The Picture Cube; **74:** UPI/Bettmann; **76:** Spencer Grant, The Picture Cube; **81:** © Jeffrey High, Image Productions; **83:** (left): Farley Andrews, The Picture Cube; (right): AP/Wideworld; **87:** Bruce Kliewe, The Picture Cube; **90:** Courtesy of the National Committee for Prevention of Child Abuse; **96:** Laima Druskis, Stock Boston

Copyright © 1995 by Globe Fearon Educational Publisher, a division of Paramount Publishing, 240 Frisch Court, Paramus, New Jersey 07652. All rights reserved. No part of this book may be reproduced or transmitted in any form or by any means, electrical or mechanical, including photocopying, recording, or by any information storage and retrieval system, without permission in writing from the publisher.

Printed in the United States of America   2 3 4 5 6 7 8 9 10   99 98 97 96 95

ISBN: 0-835-90756-2

GLOBE FEARON
EDUCATIONAL PUBLISHER
PARAMUS, NEW JERSEY

Paramount Publishing

Contents

1. **What Is Sexuality?** 1
   How Does Sex Differ from Sexuality? • Puberty and Adolescence • Making Responsible Decisions

2. **Sexuality and Your Social Life** 12
   How Does Sexuality Enter Into Your Social Life? • Are You Ready to Go Steady? • Infatuation and Love: The Differences • Different Kinds of Couples and Different Kinds of Love

3. **Reproductive Systems** 24
   The Male Reproductive System • The Female Reproductive System • What Is the Menstrual Cycle?

4. **Conception, Pregnancy, and Birth** 39
   The Creation of New Life • What Should You Know About Genetics? • Problems Relating to Pregnancy • Childbirth

5. **Preventing Pregnancy** 57
   Family Planning Is Life Planning • Barrier Methods of Birth Control • Other Methods of Birth Control • Natural Methods of Birth Control

6. **The Changing Family** 73
   Why Do We Live in Families? • Why Do People Marry? • Being a Parent • Teenage Marriage and Parenthood • What Other Options Besides Being a Parent Are Available?

7. **Sex and Society** 87
   What Are Some Unacceptable Sexual Behaviors? • What Is Meant by Sexual Abuse? • Sex and the Media

Resource List 102

Glossary 103

Index 106

CHAPTER 1
WHAT IS SEXUALITY?

John was preoccupied. His friend, Tina, asked why.

"I've been helping out in my parents' deli after school and on weekends. Business isn't that great, and my folks save money by not hiring someone else."

He continued, "I don't have time for my friends or even to do schoolwork."

Tina offered, "Why don't you just talk to your parents. Explain that you need time to do other things."

John rolled his eyes and complained, "They'll never understand. They think friends aren't as important as family."

How can John help his parents understand his dilemma? What do you think John should do?

OBJECTIVES

After completing this chapter, you will be able to:
- Explain the difference between sex and sexuality.
- Define *sex roles* and *stereotypes*.
- Describe the physical changes that take place during puberty.
- Discuss the emotional changes that occur during puberty.
- Identify some steps for making responsible decisions.

HOW DOES SEX DIFFER FROM SEXUALITY?

People often confuse the meaning of the terms **sex** and **sexuality**. Sex can refer to the act of sexual intercourse. The word, **sex**, can also refer to gender, that is, being male or female.

Sexuality is made up of all the attitudes and feelings you have about your maleness or femaleness and how you express them. Your feelings about your sexuality are intensely personal because your life experiences are unique. Your environment, religious beliefs, social customs, and what you see around you all play an important part in how you think about yourself. As you grow and develop, your ideas about sexuality will change.

Generally, attitudes about sexuality are passed from generation to generation. Parents teach about sexuality through words, by example, and by the way they treat their children. During infancy, a parent cuddles, nurses, bathes, and dresses his or her baby. The baby experiences well-being and security from the physical and emotional attention he or she receives. The baby associates parental care with acceptance.

By the age of two, a child senses that he is a boy or she is a girl and becomes curious about body parts. A parent's positive or negative reaction to curiosity, nudity, or to toilet-training is interpreted by the child either as "my body is good," or "my body is bad."

At about five years old, children begin to base their idea of sexuality on how they see their parents interact with one another. Some parents show their caring with a hug or a kiss. They teach their children that physical intimacy can be pleasurable and can be used to express love. How do you think children whose parents do not show these behaviors are affected?

What are sex roles and stereotypes?

The behaviors and expectations associated with being male or female in a community are called **sex roles**. How are sex roles learned? In most societies, our sex roles are established at birth. In infancy and throughout childhood, boys and girls are treated differently. As you probably already know, we learn some things about ourselves from the way we are treated.

Although all infants identify with their mothers, as they become toddlers, they begin to identify with the parent of the same sex. They may begin to copy gestures, sounds, and even facial expressions of the parent. As the child's circle of experience widens, he or she may begin to copy the behavior of a favorite relative or close family friend. These other adults or older siblings who are significant in a child's life may become role models. A **role model** is someone you admire and want to be like. When a child selects a role model of the same sex, the child begins to develop the sex role that will become who he or she is as an adult. Sex-role awareness is expanded during early school years when children participate in school, outside activities, and most importantly, friendships. As young people enter their teen years, they continue to imitate role models, but the circle of people they admire may be expanded to include media stars, popular classmates, or adults from outside their family.

A child who is between the ages of 15 months to 30 months is considered a toddler.

sibling: a brother or sister

It is important for people to break away from stereotypical behaviors.

In many societies, including our own, maleness and femaleness are often defined in terms of **stereotypes** (STER-ee-uh-teyps). A stereotype is the belief that all members of a group have the same characteristics. For example, in the past, girls were considered to be delicate, emotional, and dependent. They were expected to be nurturing and caring, enjoy playing with dolls, help with housekeeping chores, and eventually marry and have children. Boys were considered to be rough, physical, independent, and aggressive. It was assumed that boys would play sports, help with heavy household chores, and eventually get a job and support a family.

Women are now able to participate in activities that were traditionally male dominated.

Traditional female jobs have been in helping professions, such as teaching, nursing, flight attendance, and secretarial positions. Such jobs were seen in the past as being compatible with raising a family. At the same time, traditional male jobs have been those that have been assumed to require strength, leadership, or analytical skills. These jobs include police officer, surgeon, or scientist. Stereotypes deny the talents and uniqueness of each individual, and limit the choices and opportunities for self-development.

Popular magazines, television programs, and movies, as well as family and friends, often encourage sex roles based on these common stereotypes.

In the past three decades, there have been many changes in our society that have caused people to rethink sex roles and stereotypes. In the United States, more than one-half the women are in the work force. Men have started to share or take over the jobs of housekeeping and child-rearing. At the same time, women are sharing leadership roles in the workplace with men and are assuming positions that require physical strength, mechanical ability, or mathematical and scientific skills.

Think and Discuss

1. What is the difference between sex and sexuality?
2. How are sex roles learned?
3. How do you think male or female stereotyping can affect your own plans for the future?

PUBERTY AND ADOLESCENCE

In both males and females, development and maturation of the reproductive system is known as puberty (PYOO-bur-tee). At puberty, you are able to produce children. Generally, puberty begins between the ages of 9 and 16. In girls, its onset is usually earlier than in boys. Not everyone matures at the same rate. Puberty can last for as many as 2 or 3 years. It is marked by several dramatic physical changes.

What physical changes occur?

These physical changes include the maturing of the reproductive system and the development of secondary sex characteristics. The **endocrine** (EN-duh-krin) **system** is a group of glands located throughout the body that regulate the maturation process. These glands produce and secrete hormones. Hormones are chemicals that trigger physical changes in tissues and organs.

The **testes** are also called testicles.

Primary sex characteristics are those present at birth, the internal and external sex organs.

The beginning of puberty depends on heredity, nutrition, and overall health. In a boy, the **pituitary** (pi-TOO-uh-ter-ee) **gland** begins the process of maturation by causing the male's **testes** (TES-teez) to enlarge. This increases the amount of the male hormone, **testosterone** (tes-TAHS-tuh-rohn) in the body. Testosterone is responsible for the development of secondary sex characteristics. In the male, these characteristics include the growth of the penis, a deepening of the voice, and the growth of hair on the face, chest, underarms, and pubic area. Testosterone is also responsible for increased muscle mass and bone growth.

As with males, the pituitary gland begins the development of the female sex glands, the **ovaries** (OH-vur-eez). The ovaries increase their output of the hormone, **estrogen** (ES-truh-jen). Estrogen controls the appearance of secondary sex characteristics such as breast enlargement, the widening of hips, and the distribution of fat that shapes the female body. Another hormone causes hair growth in the underarm and pubic areas. The reproductive organs grow, and the ovaries begin the egg-production cycle that results in menstruation (men-stroo-WAY-shun).

You will learn more about the process of menstruation in Chapter 3.

SECONDARY SEX CHARACTERISTICS

Males	Females	Both sexes
pubic hair	pubic hair	skin becomes oilier
underarm hair	underarm hair	acne may develop
other body hair	breast development	increased perspiration
facial hair	menstruation	growth in height and weight
widening of shoulders and chest		

Because maturation timetables vary, teenagers tend to be self-conscious about their physical appearance. Their height, weight, skin condition, or the size and shape of various body parts often fall short of what they think is normal for their age group. It may help to know that these feelings are common among teenagers and that heredity ultimately determines physical appearance.

Adolescence (ad-uh-LES-uns) marks the end of puberty and the beginning of adulthood. It is a time of physical, sexual, and emotional changes. It is also a time of increasing independence from parents and a strengthening of a sense of self. As young people become more comfortable interacting socially, they are also learning to cope with issues of sexuality and sexual feelings.

What are the emotional changes that occur?

Adolescence can be a time of opposites: extreme sadness, then happiness; boredom, then enthusiasm; peacefulness, then anger. The same hormones that cause physical changes in the body also cause these mood swings.

Adolescence is also a time of living with tensions and conflicts that arise because teenagers are trying to break away from their parents.

For most of your life, your parents have been responsible for caring for you and for controlling your life by setting limits on your behavior. Now, you want to control such things as what you wear, what you eat, where you go, the hours you keep, and the music you listen to. You no longer want to live by your parents' rules. Your parents are not used to this. They see your changed behavior. They may hear your complaints and feel that you have become self-absorbed. You think that they are treating you like a child. You may both be right. The only way to reduce the tension between you and your parents is to talk openly, in an honest effort to understand and respect each other's needs.

As you go through this period of maturing sexually, you may get conflicting messages about acceptable ways to express your sexuality. Television and movies portray the excitement of active sexual involvement, but society strongly discourages this involvement in teenagers. It is difficult for many young people to sort out what is appropriate behavior and what is not. The result can be feelings of guilt about what they do or what they do not do.

To deal with the many conflicts of adolescence, some teenagers group together with their peers. Belonging to a group provides a teen with feelings of acceptance and strength and can provide opportunities to test newly acquired social skills. Sometimes, though, the pressure to assume the views or behaviors of the group may create a conflict. A teen who wants to belong, but who disagrees with the views of the group, may feel forced to act against his or her own thoughts, feelings, or beliefs.

Accompanying the physical changes that take place during adolescence is the maturing of the teenager's mental processes. By the end of the teen years, a person should be able to make good judgments in problem-solving situations and use logical reasoning. Through experience, both positive and negative, the older teenager should have developed a sensitivity and caring for others and should be able to make choices based on his or her own moral code.

Think and Discuss

1. **What is meant by the terms *puberty* and *adolescence*?**
2. **What are some secondary sex characteristics of males? What are some secondary sex characteristics of females?**
3. **Why do you think you and your parents sometimes have difficulty getting along?**

MAKING RESPONSIBLE DECISIONS

Suppose you have a chance to go to a party. Everybody you know is going, and you have a super outfit to wear. However, you suspect that alcohol will be served, and you don't want to drink. You begin to feel uncomfortable. You need to decide whether or not to go. You know it is illegal to serve alcoholic beverages to people under 21. If everyone is drinking, your friends will probably expect you to drink, too. You cannot be sure how some of the crowd will act if they are drinking. What if you get into a car for a ride home and find that the driver has been drinking? You understand the consequences that could result. For sure, your parents will be upset if they find out. What should you do?

Most of the decisions you make during adolescence will be difficult ones. Often your parents want to help you decide. Sometimes, you want to make your own decisions. At other times, you'll wish you could leave the decision-making to them. You may be faced with such questions as: "Should I try drugs?" "Should I have sex?" "Should I tell who cheated on the test?" "Should I continue this relationship?" Some questions will be difficult to answer because they test the values you learned at home, or as part of your religious upbringing. Some will test the values you have developed on your own. Finally, some will test the relationships you have with your peer group.

DECISION-MAKING MODEL

Step 1	Identify the issue or problem.
Step 2	Identify all the possible choices concerning the problem.
Step 3	For each choice, list the pros and cons.
Step 4	Identify the best choice and carry it out.
Step 5	Carefully review your action to make sure it was the best choice.

You can apply the decision-making model to any problem you encounter.

Peer pressure is the strong feeling a group puts on its members to be like the rest of the group. The need to feel accepted by your friends and try new things and experiences makes decision-making difficult. To be in control of your life means doing what is best for you. To know what is best, you need to think about what the consequences of your actions will be and how you will be affected. You must also consider how your decision will affect those you love.

Sometimes the use of alcohol or other drugs makes responsible decision-making impossible. People who use drugs often suffer loss of judgment, will power, and self-control. People often regret the irresponsible choices they make when they are under the influence of alcohol or other drugs.

Think and Discuss

1. What is peer pressure?
2. Apply the steps of the decision-making model to a situation of your choosing.
3. Some parents are reluctant to allow their teenagers to make important decisions. How would you convince your parents that you are capable of making an important decision?

Chapter Summary

- The term **sex** refers both to sexual intercourse and to gender. Sexuality is made up of all the feelings you have about your maleness or femaleness and how you express those feelings.
- Sex roles are behaviors and expectations associated with being male or female. A stereotype is a belief that all members of a group have the same characteristics.
- Physical changes in both sexes during puberty include increased height, maturation of the reproductive organs, and the development of secondary sex characteristics.
- During puberty, teenagers may experience mood swings. They begin to develop strong sexual feelings. They may also feel the need to become independent of parents, try new experiences, and form relationships with their peers and members of the opposite sex.
- Responsible decision-making can help you identify what you should do in certain situations.

Activities

1. A magazine has an airline advertisement that shows three male airline pilots in uniform. You are surprised because you feel a female pilot should be represented, too. Write an imaginary letter to the president of the airline. Ask for information regarding the number of male and female pilots employed by the airline. Then explain how the advertisement supports the stereotype that women are not capable of working in positions of leadership or responsibility.

2. In our society, sex roles have changed during the last three generations. Interview a parent, a grandparent, and if possible, a great-grandparent. Have them explain what behaviors were expected of them as a male or female during their teenage years and then over the course of their married years. Write a brief summary of the interview. Compare your findings with the rest of the class.

3. Think about sex-role stereotypes. Then, working with another student, examine pictures in newspapers and magazines that convey those stereotypes. Use them to make a poster of stereotypes. Write captions for the pictures that explain the stereotypes.

4. Think about a time when feelings about growing up made you happy or sad, excited or depressed, or even angry. Write a short poem that communicates how you felt and why you felt that way. You may wish to share your poem with the class.

5. Work with three other students and develop a role-play of the following situation. Suppose your friends obtained a marijuana cigarette. The joint is being passed around. Your friends watch as each person takes a drag. You don't want to smoke it, but your turn is next. What would you do? In your role-play, suggest strategies that could help you handle the situation.

Thinking It Over

1. Joan collected her books after a study session in the school library. As she stepped out into the hallway, she saw Steve quietly sitting at a chess table. She had been wanting to get to know him better, but he never seemed to be alone. After a little small talk, Steve asked her if she played chess. He invited her to play, so she began setting up her chess pieces. It wasn't long before Joan realized that Steve couldn't play well. She began to worry how he would feel if she won the game. Would he feel embarrassed losing to a girl? If she won, would he ever ask her out? Should she let him win?

   What would you do if you were Joan? Choose the option below that you think is best, and explain why you chose that option.

   Options:

   a. Make a wrong move or two and let him win.

   b. Play the game to win.

   c. Play the game, pointing out Steve's wrong moves and giving him the chance to correct them.

   d. Lose the game on purpose. Then ask him to give you chess lessons.

2. You and your best friend are trying out for the gymnastics team. When the lists are posted, you see that you were chosen and your friend was not. You notice that two other gymnasts who qualified had not performed as well as your friend. Without telling your friend, you ask the team captains to explain. They tell you that your friend would not have fit into the group socially. "If you want to join the team," they say, "forget your friend." What would you do in this situation? Explain your reasons.

CHAPTER 2
SEXUALITY AND YOUR SOCIAL LIFE

Lionel and Joe have stood by each other through thick and thin–algebra, spring training, being dumped by their girlfriends the day after the prom. This afternoon, Joe is acting weird. Lionel asks him what's wrong.

Joe takes a deep breath. "My mother's dating a woman!" Joe starts to say more, but breaks off when he sees the look on Lionel's face.

Joe is looking more upset by the second. Lionel knows that Joe needs his support, but he does not know how to respond. What would you do?

OBJECTIVES

After completing this chapter, you will be able to:
- Explain why it is important to choose carefully the peer groups to which you belong.
- Identify the advantages and disadvantages of going steady.
- Distinguish between infatuation and love.
- Outline an effective strategy for resisting peer pressure.

HOW DOES SEXUALITY ENTER INTO YOUR SOCIAL LIFE?

Have you ever overheard a teacher or other adult expressing concern about a teenager who fell in with the "wrong crowd"? Why was the adult concerned? How do you think falling in with the "wrong crowd" affects a teenager's decisions about sex or dating?

Try to identify four or more crowds, or peer groups, in your school. How would you describe the various groups that you identified? Perhaps one group attracts people who have similar interests, such as sports, reading, or fashion. Perhaps another group attracts people with similar family situations or religious backgrounds.

Each group undoubtedly has its own expectations regarding the sexuality of its members. Do you know what those expectations are? Suppose that you want to postpone having sexual intercourse until you are married. Which of the groups that you identified could help you achieve this aim? Which of these groups would it be best to avoid?

In answering the questions above, you may have realized that groups exert a stronger influence on individuals than individuals exert on groups. The groups to which you belong as a teenager will play a big part in determining not only how and when sexuality enters your social life, but also the kind of adult that you become. That's why adults worry about teenagers falling in with the "wrong crowd." To become the best adult that you can be, it is important to choose your friends wisely.

How can you tell if the group to which you belong is emotionally healthy? Does your group provide support for its members, while respecting the rights of individuals to make their own decisions? If not, you may need to examine why you belong to such a group and why the members of the group are unable to interact in a way that strengthens each individual.

Sexuality and friendship

As you know, not every teenager hangs out in a crowd, or even spends time with a small group of friends. There are many reasons such as family obligations or after-school jobs, that can prevent teenagers from participating in before- or after-school activities. Almost all teenagers, however, from the outgoing and popular to the shy and reserved, have at least one close friend.

What do you look for in a friend? What do you share with him or her? Do you share experiences? Information? Possessions? Perhaps like many teenagers, you are inseparable from your closest friend. You save seats for each other in the lunchroom, spend afternoons and weekends together, and even go on vacations with each other's family.

Unfortunately, close friends of the same sex sometimes become the targets of gossip. Rumors circulate that so-and-so is a **homosexual** (hoh-muh-SEK-shoo-wul), or a **bisexual**. Is it true that close friends of the same sex are homosexual or bisexual? In most cases, it is probably not true. Just as you can enjoy a close emotional bond with parents or siblings, you can also experience deep, **platonic** bonds of affection with your friends. The depth and quality of teenagers' friendships reveal more about their capacity to give and receive affection than about their sexual orientation.

Stereotypes about homosexuals, like stereotypes about racial, ethnic, and religious minorities, are untrue. The football captain who is **heterosexual** (het-ur-uh-SEK-shoo-wul) can have many close male friends. In contrast, a teenage lesbian may be so afraid of being "discovered" and rejected by her peers that she actually has very few female friends.

The vast majority of friendships between teenagers are platonic. Yet, as you may know from your own experience, platonic friendships between teenagers of the opposite sex can become romantic.

homosexuals: people who are sexually attracted to members of the same sex.

Women who are homosexuals are called lesbians.

bisexuals: people who are sexually attracted to members of both the same and the opposite sex.

platonic friendship: emotional bond between two people who are not sexually attracted to each other.

heterosexuals: people who are sexually attracted to members of the opposite sex.

Friends of the same sex can offer support and advice.

Think and Discuss

1. Describe the function of groups in a teenager's life.
2. What is a platonic friendship?
3. Tenicia and Maria are always huddling together and whispering secrets. Rumors are circulating about Tenicia and Maria. You are friends with both girls. How do you deal with the rumors?

ARE YOU READY TO GO STEADY?

What words come to mind when you think about dating? Competitive? Fun? Expensive? Sexual? How about educational?

Education may be the last thing on your mind when you think of dating, especially if you think of education as only book learning! Yet, dating and learning are closely related. Consider the following situations.

Teenagers are attracted to each other because of such factors as personality, behavior, experiences, interests, and appearance.

You have dinner at the home of a date whose parents are very formal. At your place setting, there are eating utensils that you have never seen. You have no idea which fork to use first, or which salad plate is yours. You care a lot about your date, however, and are determined to conduct yourself well. You pay close attention to how others at the table are eating. You feel a little nervous, doing as they do, but you appear perfectly at ease. At the end of the evening, you feel proud of yourself. Moreover, in successfully negotiating a more formal dinner than occurs in your home, you have attained new skills that you can use in other social settings.

Sometimes dating teaches more painful lessons. Suppose, for example, that the most popular boy or girl in school chooses you as a date for the prom. You are ecstatic. You spend all the money that you have earned in the last three months on the prom. The most popular student in school, however, behaves like a real fool. He or she does not listen when you talk, makes nasty comments about your friends, and constantly talks about the other people he or she has dated in the past. Your prom is ruined, yet you have learned something. You now recognize that popularity does not automatically mean that a person is sensitive or even has social skills. You can use this knowledge when you need to make decisions about other people you may want to date.

Dating allows teenagers to learn from experience. Dating a variety of people can help you identify the qualities in a person that you feel are essential for a healthy relationship. Dating can also broaden and enrich your life by allowing you to interact with people who come from backgrounds different from yours.

Going steady

Eventually, you will be captivated by one particular person, so much so that you have no interest in dating anyone else. If that special person asks you to go steady, your instinct will be to answer "Yes!" You will be better equipped to decide what is best for you, however, if before receiving an invitation to go steady, you know the pros and cons.

What advantages do steady couples have? Steady couples do not have to worry about sitting at home on Saturday nights, or not having a date for a big dance or party. Steady couples are able, through the time that they spend together, to develop emotional intimacy. Teenagers who go steady experience firsthand the excitement and joy that comes from knowing that the person they care about most feels the same way about them. Finally, steady couples can acquire skills and confidence in dealing with stresses, such as jealousies and misunderstandings, that occur in a long-term relationship such as marriage.

There is, however, a downside to going steady. By involving themselves exclusively in one dating relationship, steady couples limit their opportunities to interact with a variety of people. Some teenagers who have not had much dating experience may accept whatever happens in their steady relationship. This could include emotionally abusive, or even violent behavior. Finally, going steady intensifies all the aspects of a dating relationship, including sexual desire. You need to have a clear sense of, and appreciation for, who you are before you make a commitment to date only one person.

Think and Discuss

1. What can you learn through dating?
2. Name two advantages and two disadvantages of going steady.
3. Eduardo is afraid that Anna, the only girl he wants to date, is interested in someone else. To find out where he stands with Anna, Eduardo plans to ask her to go steady. Do you think that this is a good plan? Explain.

INFATUATION AND LOVE: THE DIFFERENCES

Do you ever play charades? Think about how the game begins. People take turns acting out, or communicating without words, the title of a book or movie. Sometimes charade players start by pulling their ear lobes, pointing toward their eyes, or poking themselves in the ribs. "Sounds like! Looks like! Feels like!" the other players respond.

Infatuation (in-fach-oo-WAY-shun) is an emotional charade. It feels like love. It looks like love. But it is not love.

How can you tell if you are infatuated or in love? One way is by considering realistically your chances of building a relationship with the person to whom you are attracted.

A second way to differentiate infatuation and love is by asking yourself how well you know the other person. You can become infatuated with almost anyone, including a movie star, a political leader, or the older sibling of a friend. Love is more selective. You cannot love someone whom you do not know and who does not know you. Infatuation feeds off fantasies of what a relationship can be. Love is nourished by the reality of two people coming together in mutual caring, sharing, and understanding.

Suppose that you feel surges of attraction toward someone whom you do know well, perhaps a platonic friend of the oppo-

infatuation: short-lived, yet intense attraction toward a person whom one does not know well.

site sex. Is it infatuation, or is it love? The best way to tell is by observing your feelings over time. Infatuation wears off as you learn more about who the other person really is. Love grows as the other person enters your reality, as well as your dreams. When people in love come to know each other well, they are more likely to want to share their lives.

The strong similarities between love and infatuation support a wait-and-see approach before acting on feelings of sexual attraction. You can enjoy the good feelings of an infatuation while holding out for love.

How can you resist peer pressure?

Suppose that the test of time proves your love is "for real." As the school year progresses, you and your partner develop a relationship that reflects mutual concern, sharing, and affection. How will your sexuality enter into this relationship? Does it follow that since you are in love, you should go "all the way?"

Many of your peers will say "yes." Your boyfriend or girlfriend may agree. Suppose that you, like many teenagers, do not feel ready for sexual intercourse. What do you say to your boyfriend or girlfriend? What do you say when the subject arises among your peers?

First of all, be aware that everybody is not doing *it*, regardless of what *everybody* says. Second, remind yourself that the decision about when and with whom you should become sexually active is not a group decision. You are the one who must be comfortable with the risks and consequences of whatever you decide.

Peer pressure to become sexually active is easiest to address when it comes from platonic friends. If your friends try to force their opinions on you, point out, politely but firmly, that your sexuality is none of their business. You can cut off a friend's questions by saying, "That's between (name of boyfriend or girlfriend) and me."

Effectively resisting peer pressure takes practice. Some teenagers find it helpful to create imaginary pressure situations and formulate responses to them. You might try observing yourself in a mirror, or working with an adult whom you like and trust. Try to develop a way of responding that, without putting anybody down, clearly communicates how you feel. Stand up straight. Make eye contact. Try to look calm, confident, and relaxed—not nervous or tense.

Pressure to become sexually involved can be very difficult to resist when it comes from someone whom you love. Suppose that your steady boyfriend or girlfriend is pressuring you to go farther than you want. You know what you want to communicate, but you have little confidence that the words will come out right. Or you may have mixed feelings about the situation. You may feel unready for sex but deeply attracted to your boyfriend or girlfriend. So you are uncertain of what you want to say.

Talking honestly about feelings is a skill young people need to learn.

Do not worry if you feel uncomfortable discussing sexuality with your girlfriend or boyfriend. Adults who have been married for years often have trouble discussing sexual issues and concerns. You can increase the likelihood that your boyfriend or girlfriend will accept and understand your point of view, if you remember that his or her feelings are deeply involved.

Try never to assert sexual limits in a manner that implies that you do not find your boyfriend or girlfriend attractive. Use "I" statements to express what you want or do not want. For example, say, "I am not ready for sexual intercourse," rather than "Good people wait until marriage." The former is a statement about you. The latter implies that there is something wrong with someone else.

Think and Discuss

1. Name one way in which infatuation and love are alike and two ways in which they are different.
2. What are some guidelines for responding effectively to peer pressure?
3. Lanita and Santiago have been going steady for 10 months. Lanita wants a sexual relationship, but Santiago does not feel ready. "Don't you find me attractive?" Lanita asks. What can Santiago say?

DIFFERENT KINDS OF COUPLES AND DIFFERENT KINDS OF LOVE

As a teenager, most of the couples that you observe are heterosexual. As you grow in age and experience, you will have acquaintances and, perhaps, friends involved in homosexual relationships. You may become aware of couples in which one or both partners are bisexual. You may discover that your own sexual orientation is different from that of the heterosexual majority.

As your familiarity with different types of couples grows, you will become aware that real-life couples, whatever their sexual orientation, tend to have much more in common than media portrayals lead people to believe.

In fact, like married heterosexuals, two homosexuals may live in a monogamous (muh-NAHG-uh-mus) relationship in a family that includes a child whom they have adopted. Similarly, a bisexual man or woman may live in a monogamous, heterosexual marriage.

monogamous: emotionally and sexually committed to one other person.

DOES THE GROUP TO WHICH YOU BELONG...

- Feel comfortable to you?
- Have many of the same values that you have?
- Support your right to make your own choices?
- Respect you for who you are, instead of pressuring you to change?

Think and Discuss

1. How can media portrayals affect peoples' perceptions of same-sex relationships?
2. Name two similarities that can occur in same-sex and heterosexual relationships.
3. Joe's mother and her girlfriend, along with his father, plan to attend Joe's graduation. Joe is worried about how to introduce his mother's girlfriend to his friends. Give Joe a suggestion.

Chapter Summary

- The peer groups to which you belong as a teenager will strongly affect the kind of adult that you become.
- It is important to consider the pros and cons, as well as your emotional readiness, before going steady.
- Infatuation and love bear a strong resemblance to each other, but they are not the same.
- Through practice, you can learn to effectively resist peer pressure.

Activities

1. Brainstorming is a technique used to collect many ideas on a topic. Work with three other classmates to brainstorm a list of desirable qualities in a platonic friend. Remember, when you brainstorm, do not criticize each other's ideas. After your group has completed the list, have one group member report to the rest of the class.

2. Interviewing is a technique used to find out people's opinions, ideas, or knowledge about a topic. Write down four questions about going steady. Interview at least three teenagers. Compare the responses you receive. What are the similarities and differences? Compare your responses with those of the other members of your class. Were there any similarities in the responses they received? Summarize your findings in a brief written report.

3. Work with three other students to develop and present a skit on infatuation versus love. In the skit, show one couple that is infatuated and one couple that is in love. Have the class identify which couple is which. Have students explain how they could tell the difference between the infatuated couple and the couple in love.

4. With one other student, develop a role-play showing effective ways to resist pressure to have sexual intercourse. Demonstrate how you would say "no" to someone pressuring you to have sexual intercourse. Then switch roles and play the other part in the role-play. How did you feel playing each of the roles?

5. Go to your school or local library and research how heterosexuals, homosexuals, and bisexuals are commonly portrayed. A computer index such as InfoTrak can help you locate materials. Find three or four articles. How are people of different sexual orientations portrayed? Write a brief summary to present along with your articles. Include in your summary the titles of the articles, as well as the title of the magazine or newspaper from which they came.

Thinking It Over

1. Cheryl's social life was difficult until this year, when she became a member of the most popular group in school. Cheryl is thrilled. She no longer worries about where to sit in the lunchroom or having a date on Saturday nights. Most of the teenagers in Cheryl's new group are going steady. Several have become sexually active. Although she is dating someone regularly, Cheryl does not feel ready for either a steady boyfriend or a sexual relationship.

   What would you do if you were Cheryl? Choose the option below that you think is best, and explain why you chose that option.

Options

   a. Lie about your sexual activity.

   b. Accept your boyfriend's invitation to go steady, but tell him that it doesn't really mean anything.

   c. Tell yourself that it's okay to have sexual intercourse, as long as you use birth control.

   d. Develop relationships outside the group.

   e. Find another group of friends.

2. Imagine that a boy in your class is regularly being beaten up after school because he is homosexual. Even students who do not participate in the physical violence refer to this boy as a "queer" or a "fag." What would you say or do if the boy was victimized, physically or verbally, in your presence? Write what you would do or say.

CHAPTER 3
REPRODUCTIVE SYSTEMS

When Aaron and Luisa met before school, she had a worried look on her face. "What's wrong?" asked Aaron. Luisa looked around to make sure no one was listening, then whispered, "I'm pregnant."

"What?" Aaron sputtered. "Are you sure?"

"Yes. I went to the doctor."

"But how could it have happened?" wondered Aaron.

"Well, you know we didn't use birth control the time we had sex," Luisa gulped.

"I'm still sure I pulled out in time!"

Aaron and Luisa thought they were being careful enough to avoid pregnancy. They were wrong. What do you think they didn't know?

OBJECTIVES

After completing this chapter, you will be able to:

- Utilize proper terminology for each of the sexual organs.
- Name and describe the function of each male sex organ.
- Explain how sperm is released from a male's body.
- Name and describe the function of each female sex organ.
- Understand the process of menstruation.
- Examine either your breasts or your testicles for signs of cancer.

THE MALE REPRODUCTIVE SYSTEM

The biological purpose of a man's reproductive system is to produce **sperm**. It also must deliver the sperm to the female's reproductive system. Sperm are tiny cells. They can be seen only through a microscope. A sperm cell is shaped like a tadpole. It has an oval head and a long, slender tail used for swimming. Each sperm carries genetic material. When a sperm joins with an egg produced inside a female's body, the genetic materials of the two cells are combined. This process, called **fertilization** (fur-tul-i-ZAY-shun), enables the egg to begin developing into a human being.

Sperm are produced by two oval-shaped organs called testes. The testes lie side-by-side inside a pouch called the **scrotum** (SKROHT-um), which hangs behind and below the penis. The scrotum is outside the body because the production of sperm requires a temperature a few degrees cooler than normal body temperature. Depending on the temperature outside the body, muscles in the scrotum will either contract or relax. When the muscles contract, they draw the testes towards the body to warm them. When the muscles relax, they bring the testes farther away from the body to cool them. A man may be aware of this after coming out of a hot shower, or when dressing in a cold room.

After sperm are produced, they are stored in the **epididymis** (ep-uh-DID-i-mus). One of these storage organs curls around the back of each testicle.

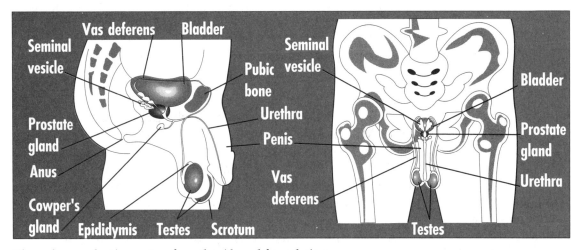

The male reproductive system from the side and frontal view.

How are sperm released from the body?

The release of sperm from the penis, called **ejaculation** (i-jak-yuh-LAY-shun), is part of a process that involves several other male sex organs. The process begins with sexual excitement, which may be brought on by different kinds of sensory experiences, such as looking at a picture or touching someone you're attracted to. Sexual excitement can cause spongy tissues in the penis to fill with blood, making the penis larger, longer, and harder. This change is called an **erection**. An erection makes it possible for the penis to be inserted into a woman's vagina.

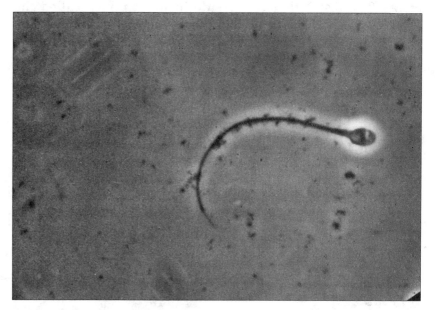

Millions of sperm are released in semen each time a male ejaculates.

The **urethra** also carries urine out of the body. During a period of sexual excitement, muscles prevent urine from entering it.

If the erect penis is stimulated, the process enters another stage. Millions of sperm begin to move out of the epididymis through a pair of tubes called the **vas deferens** (DEF-uh-renz). Each of the vas deferens is connected to a testicle. These tubes loop past the bladder and join the **urethra** (yoo-REE-thruh). The urethra is a tube that leads out of the end of the penis.

As the sperm pass through the vas deferens, fluids are added that nourish and protect them. A large part of this fluid comes from the **seminal vesicles** (SEM-uh-nul VES-i-kuls). The seminal vesicles are glands that are located under the bladder. More fluid is added by the prostate gland, where the vas deferens joins the urethra, and by Cowper's glands. The mixture of the sperm and the fluids is called **semen** (SEE-mun).

If sexual stimulation continues, there comes a moment when muscles around the vas deferens and urethra begin to contract rhythmically, causing ejaculation. The contractions, which continue for about 10 seconds, propel the semen contained in the tubes through the penis and out of the body. The muscle contractions are combined with feelings of intense pleasure called **orgasm**. It is important to know when you become sexually active, that fluid containing sperm can be released from the penis even before ejaculation occurs.

Is what you're experiencing normal?

During adolescence, many males have worries about their sexual organs. Sometimes, for example, boys who have reached puberty have orgasms and ejaculate in their sleep. This sexual experience is a **nocturnal emission**. Nocturnal emissions are often called wet dreams. It is normal to have nocturnal emissions from time to time. Many boys also have erections for no apparent reason, even when they are not having sexual thoughts. These are called spontaneous erections. Although these sudden erections may be embarrassing, they, too, are normal.

Many males and females learn to **masturbate**, or stimulate their own sexual organs for pleasure. This is another source of concern because much shame and guilt surrounds masturbation in our society. Parents may punish a child they find masturbating and may give out false information about how masturbation can be harmful. Experts agree, however, that masturbation is a completely healthy and normal way to experience your sexuality. In fact, mutual masturbation is often recommended as a safe alternative to sexual intercourse.

Many boys also worry about the size of their penises. There is natural variation in penis size and shape, but during adolescence much of it is due to different timetables of maturation. Also, you should know that penis size has little to do with the pleasure experienced by either a man or his partner during intercourse.

Some penises have a flap of tissue covering the head. Others do not. At birth, this flap of tissue, called the foreskin, covers the head of the penis. The foreskin may be removed shortly after birth during a simple surgical procedure called **circumcision** (sur-kum-SIZH-un). Circumcision has been part of a religious practice among Jews and Muslims for many centuries. In the United States, many people have their boy babies circumcised for hygienic reasons alone. A waxy substance is secreted by the tip of the penis. If a male is not circumcised, this substance will collect under the foreskin. To prevent odor or infection caused by this, uncircumcised males should pull back the foreskin when bathing or showering to clean the head of the penis. Whether or not a man is circumcised makes no difference in the way he functions sexually.

Why should men examine their testicles?

It is possible for young men to develop cancer in one of their testicles. In fact, testicular cancer is the most common kind of cancer in young men. It can be cured if it is discovered in its early stages.

cancer: a disease caused by the uncontrolled growth of a group of cells called a tumor.

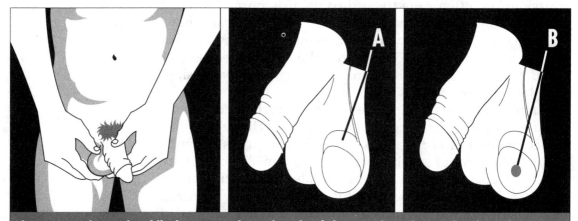

Place your index and middle fingers on the underside of the testicle and your thumb on the top. Gently roll your testicle between your thumb and fingers. The normal testicle feels slightly soft with an even consistency and a smooth surface. The epididymis (A) can be felt at the back of the testicle and feels slightly different in consistency. Examine both testicles. Any thickening or lump (B) however small should be reported to your physican.

All men should do a testicular self-exam regularly.

A doctor will examine a boy's or man's testicles for cancer as part of a regular physical exam. These exams, however, don't always occur often enough to detect testicular cancer in time. For this reason, boys and men should examine their testicles every month for signs of cancer. Doing a testicular self-examination is easier after a hot shower, when the testes are most relaxed. Using both hands, be sure to check each testicle separately. Feel for pea-sized lumps or hard places by gently rolling the testicle between your thumb and two fingers. If you find what you think is a lump, see your doctor as soon as possible. Check with a doctor, too, if you notice an enlargement of one testicle over a period of time.

*Before doing a testicular exam, check to be sure you know where the **epididymis** is located so you do not mistakenly think it is a lump.*

Think and Discuss

1. What is a sperm cell, and what purpose does it have?
2. Describe the route sperm travel after they leave the testes.
3. During a simple medical procedure called a vasectomy, the vas deferens is cut. What do you think is the purpose of a vasectomy?

THE FEMALE REPRODUCTIVE SYSTEM

For the purpose of reproduction, a woman's reproductive system has a much more difficult job than a man's. In addition to producing **eggs**, or ova, the female sex organs have other functions. They receive sperm and help them reach an egg in order to fertilize it. In addition, some of the female sex organs nourish the fertilized egg as it grows from a single cell into a complete human being.

What are the organs of the female reproductive system?

A pair of plum-sized organs in a woman's lower abdomen, called ovaries, perform two main functions. They produce hormones that create sexual feelings and control the reproductive system, and they produce eggs. Ovaries function in much the same way as the male's testes in that they, too, make sex cells.

hormones: substances produced by various organs and glands in the body that regulate body processes.

All the eggs a woman will ever have are already in her ovaries when she is born. These eggs remain undeveloped inside the ovaries until puberty. At puberty, some eggs near the edge of the ovaries begin to become mature. Each egg, or ovum, grows inside an envelope of cells called the follicle (FAHL-i-kul).

Only 1 egg matures at a time. When an egg is fully mature, its follicle—now about the size of a pea—breaks open and releases the egg. This process, which occurs about once every 28 days, is called **ovulation** (oh-vyuh-LAY-shun).

Fallopian tubes are sometimes called oviducts.

When an egg is released, it enters a **Fallopian** (fuh-LOH-pee-un) **tube.** There are two Fallopian tubes, one leading from each ovary. An egg is fertilized as it moves down one Fallopian tube or the other. This can occur if a woman has had intercourse and sperm reach the egg when it is passing through the Fallopian tube.

The uterus is one of the female organs most prone to the development of cancer. Starting at age 25, a woman should have a yearly Pap Test. A Pap Test is used to detect any abnormal cell growth.

The Fallopian tubes connect to the **uterus** (YOOT-ur-us). The uterus is a muscular, pear-shaped organ that is also called the womb. If an egg is fertilized, it will attach to the wall of the uterus, and the uterus will nourish it as it develops.

At the lower end of the uterus is the **cervix** (SUHR-viks). The cervix has a small opening that can enlarge greatly during childbirth to let the baby leave the uterus. The cervix leads into the **vagina** (vuh-JI-nuh). The vagina is the passageway from the uterus to the outside of the body. The vagina is both a birth canal and the place where sperm is deposited by the penis during intercourse.

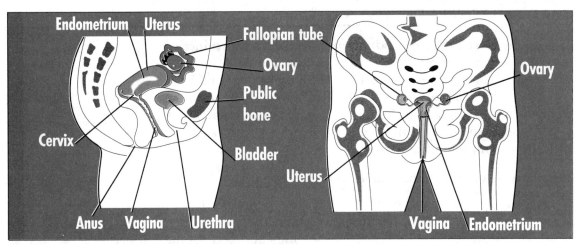

The female reproductive system from the side and frontal view.

The opening of the vagina may be partially covered by a membrane called the **hymen** (HI-mun). Often the hymen is broken by exercise or accidents during childhood, or by the insertion of tampons. If this does not occur, it may be broken or stretched the first time a girl or woman has sexual intercourse.

The opening of the vagina is surrounded by two sets of skin folds called **labia** (LAY-bee-uh). Labia are shaped something like lips. The outer labia are covered by pubic hair; the smaller inner labia are moist and very sensitive to touch. Within the inner labia and above the vaginal opening is the **clitoris** (KLIT-ur-us). The clitoris is a small organ that contains many nerve endings and blood vessels that fill during sexual arousal. In this way, the clitoris functions much like a male's penis. Between the clitoris and the vaginal opening is the opening of the urethra, through which urine leaves the body.

When a woman becomes sexually aroused, her clitoris becomes swollen, and the walls of her vagina produce mucus. Stimulation of the clitoris, labia, and vaginal walls, whether through sex with a partner or masturbation, can lead to orgasm. Like the male orgasm, a woman's orgasm involves muscular contractions and intense pleasure. A woman's orgasm may last longer than a man's and may be repeated several times.

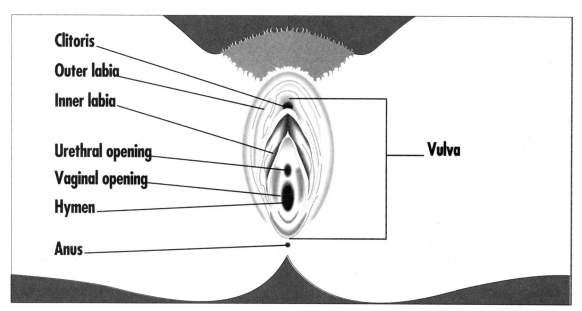

The external organs of the female reproductive system.

Think and Discuss

1. Define *ovulation*.
2. What are the two functions of the ovaries?
3. Since the clitoris is covered by the labia, some women are unsure of its exact location. What can someone who is unsure about where her clitoris is do to locate it?

WHAT IS THE MENSTRUAL CYCLE?

Recall that ovulation occurs about once every 28 days. The regular intervals between ovulation set up a cycle in the functioning of the female reproductive system. During each cycle, a woman's body goes through many changes as it releases an egg, enabling the woman to become pregnant. If a pregnancy doesn't occur, the body changes again as it gets ready to restart the cycle. This is very different from what occurs in the male reproductive system, which produces sperm, not monthly, but continually.

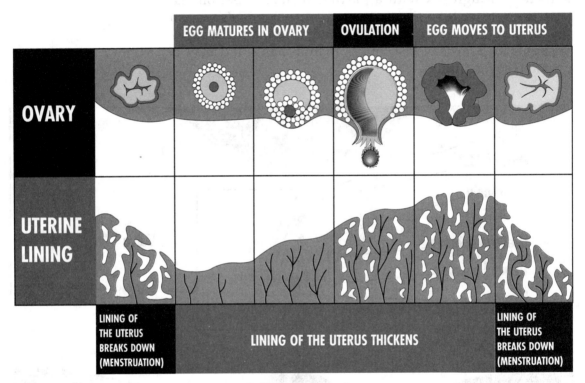

The process the body goes through each month during menstration.

During a woman's menstrual cycle, hormones coordinate what happens in the ovaries and in the uterus. As an ovary gets ready to release a mature egg, the lining of the uterus begins to thicken. The lining of the uterus is called the **endometrium** (en-doh-MEE-tree-um). It builds up a rich supply of blood vessels that will help it nourish an egg if a fertilized one arrives in the uterus.

After ovulation occurs, the follicle, now called the **corpus luteum** (KOWR-pus LOO-tee-um), produces the hormones

progesterone (proh-JES-tuh-rohn) and estrogen. Progesterone keeps the endometrium thick and causes it to increase its supply of blood vessels. The corpus luteum continues to produce progesterone while the egg passes through the Fallopian tube and into the uterus—a period of about seven days.

If the egg has not been fertilized by the time it reaches the uterus, the corpus luteum stops producing progesterone. Without this hormone, the endometrium begins to break down. Its blood vessels collapse and its tissues die. The endometrial matter is shed from the uterus and discharged from the body along with the unfertilized egg, blood, and other fluids. This process is called **menstruation**. After the menstrual flow stops, the cycle repeats itself as another egg begins to mature in an ovary.

When during her menstrual cycle is a woman fertile?

A woman is most fertile, or able to become pregnant, when she is ovulating and during the few days following ovulation. If she has sexual intercourse during this period, the sperm will have time to travel through the cervix, the uterus, and into the Fallopian tubes. Here the sperm intercept the egg and fertilize it.

Although the period of fertility during each cycle is relatively short, it is not completely predictable. Ovulation may not always occur the same number of days after the menstrual flow stops. It can happen earlier than expected or be delayed for days or even weeks, especially in girls and young women. Moreover, sperm can sometimes live a long time in a woman's body. If a woman has sexual intercourse before she ovulates, live sperm may still be in the Fallopian tubes. When the egg passes through, these sperm can impregnate her.

How do menstrual cycles vary?

Girls experience their first menstruation, or period, at different ages. Menstruation can begin any time between ages 8 and 16. For most girls, it happens between ages 10 and 14. At first, a girl's cycle may be irregular, often with large variations in the amount of discharge and the spacing between periods. But as a girl matures, her cycle becomes more regular and predictable.

Every woman's menstrual cycle is different, and there is a wide range in what is considered normal. Although the average time from the beginning of 1 cycle to the beginning of the next is 28 days, any length from 25 to 40 days is normal. The period of menstrual flow can last anywhere from 3 to 7 days, and the amount of menstrual flow, or discharge, can vary from heavy to light.

Many individual women find that their own cycles can vary, too. Changes in diet, great amounts of stress or heavy exercise, illness, and significant weight gain or loss can all affect a woman's cycle. The length of her cycle can change, she can experience more discomfort or heavier discharge than usual, or she can skip a period altogether.

During the several days before they begin to menstruate, some women experience a variety of physical and emotional changes. They may cry easily, feel irritable or depressed, or feel bloated, or full of water. These symptoms are part of what is called **premenstrual syndrome**, or **PMS**. Premenstrual syndrome is normal. For some women, however, its symptoms may be severe enough to interfere with day-to-day living.

Most women continue menstruating and ovulating until sometime between the ages of 45 and 55. At that time, a woman's menstrual cycle becomes irregular and then stops. This stage of life is called **menopause** (MEN-uh-pawz). After menopause, a woman is no longer fertile, but she can still have a full and satisfying sex life.

Why should women examine their breasts?

While a woman's breasts are not part of her reproductive system, they do have a reproductive function: providing milk for an infant. Unfortunately, the breasts are susceptible to developing cancer. As a result, breast cancer is one of the most common kinds of cancer among women. As a woman grows older, her chances of developing breast cancer increase. The cause of breast cancer is unknown, but women who have a family history of the disease are at higher risk than others.

If breast cancer is detected in its early stages, it can be treated and often cured. For this reason, doctors advise all women to examine their breasts regularly for signs of the disease. The cancerous cells form a hard lump that feels different than normal breast tissue. If a girl or woman performs breast self-examinations monthly, she learns what all parts of her breasts normally feel like. This makes it easier to detect an early-stage cancer lump. The best time to do a breast self-exam is two weeks after the start of the last menstrual period.

If you feel anything unusual during a breast self-exam, you should see a doctor. There is no need to panic, however, because harmless lumps in the breast are common. Only a doctor can determine for sure if a lump is cancerous.

In the shower: Raise one arm. With fingers flat, touch every part of each breast, gently feeling for a lump or thickening. Use your right hand to examine your left breast, and your left hand for your right breast.

Before a mirror: With arms at your sides, then raised above your head, look carefully for changes in the size, shape, and contour of each breast. Look for puckering, dimpling, or changes in skin texture.

Then, place your hands on your hips and look again carefully.

Lying down: Place a towel or a pillow under your right shoulder and your right hand behind your head. Examine your right breast with your left hand.

Keep your fingers flat. Press gently in small circles, starting at the outermost top edge of your breast, and circling in toward the nipple. Examine every part of your breast. Repeat with the left breast.

Lastly, gently squeeze the nipple of each breast between the thumb and index finger and look for discharge. Any discharge or lumps should be reported to a doctor right away.

Think and Discuss

1. What is being removed from a woman's body during menstruation?
2. Why is it important for a woman to do a breast self-exam regularly?
3. Shawna is sexually active. It has been almost six weeks since her last period, and she is convinced that she is pregnant. How would you explain to Shawna that she might not be pregnant?

Chapter Summary

- The male reproductive system is designed to produce and deliver sperm. Sperm are produced in the testes. The vas deferens and urethra are tubes that carry the sperm to the penis. The seminal vesicles, prostrate gland, and Cowper's glands produce fluid that mixes with sperm to create semen.
- Semen is released from the penis by muscle contractions during ejaculation.
- The female reproductive system is designed to produce eggs and nourish a developing fetus. Eggs are produced in the ovaries. They travel through the Fallopian tubes to the uterus. If an egg is fertilized, it stays in the uterus where it develops.
- When an egg is released from an ovary and is not fertilized, it is discharged from the body with the endometrium as part of the process of menstruation.
- To detect testicular cancer and breast cancer at their earliest stages, it is important for men to examine their testicles regularly and for women to examine their breasts regularly.

Activities

1. Interview two or three teenage and/or adult women. Ask them to describe how they usually feel just before menstruating. How many of the women suffer from premenstrual syndrome? What other variations do you find in their experiences?
2. Work with four other students and design a poster that shows the organs of the male and female reproductive systems. Label each organ and describe its function.
3. In the male, cancer can occur in the prostate gland, as well as in the testes. Research prostate cancer, and write a short report in which you answer these questions: What are the symptoms of prostate cancer? How can it be detected in its early stages? How is the disease treated? How common is prostrate cancer in men of different age groups?
4. In the female, several of the reproductive organs, including the uterus, ovaries, and the cervix, may become cancer sites. Work with two other students and find out more about these types of cancers. Prepare an oral presentation of your findings to the class.
5. Form a group of girls only and a group of boys only. In each group, think of questions you would like to ask about the bodies or reproductive systems of the opposite sex. Have a volunteer write down all the questions on a single sheet of paper, leaving

space below each one. Then trade questions with the other group. Within your group, read each of the other group's questions and answer them in writing as best you can. When you're done, return the completed question sheets. Read and discuss the answers you get to your questions.

Thinking It Over

1. Jeremy has just moved and is in a new school for the start of eighth grade. As he gets ready for gym class, he notices that he is one of the few boys without any pubic hair and that his penis seems much smaller than those of the other boys. The other boys notice, too, and make fun of him. What should Jeremy do? Choose the option below you think is best, and explain why you chose that option.

    Options

    a. Do his best to avoid having the other boys see him naked.

    b. Tell the boys that his father's penis was small at his age and now that is no longer the case.

    c. Just accept the criticism and walk away.

    d. Realize that he's just on a different "timetable" than the others and that eventually, he will start to grow.

2. Pam does her first breast self-exam and feels what seem like several lumps. Should she go see a doctor immediately, or just wait and see if the lumps change? Give reasons for your answer.

3. Imagine that your 14-year-old sister, Rowana, has confided in you that she still hasn't begun to get her period. Her friends have all gotten their periods. Whenever she is with her friends, it seems to her as if that's all they talk about. Finally, Rowana told her friends that she, too, had gotten her period. Some of her friends said they didn't believe her. To convince them, Rowana wants you to come into her room the next time her friends are over and hand her a box of tampons. Should you do it? What would you say to her?

CHAPTER 4
CONCEPTION, PREGNANCY, AND BIRTH

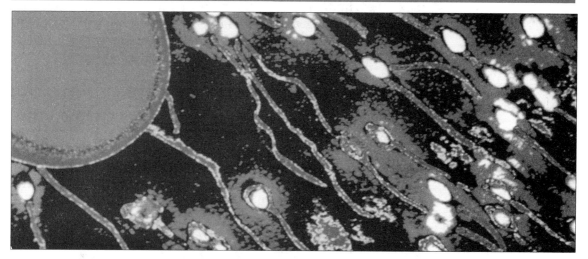

Vera was beginning her eighth month of pregnancy. She just had an ultrasound examination and was worried about what she had found out.

She decided to call her mother. "Mom, I'm really scared about what the ultrasound showed." Vera's eyes began to moisten. "It's a breech baby—it's upside down. The baby's *feet* are going to be born first!"

Vera's mother told her not to worry.

"But Mom, they said it's dangerous, and I'm afraid the baby is going to die!"

Who is right? What do you think?

OBJECTIVES

After completing this chapter, you will be able to:

- Explain the process of fertilization.
- Describe how a fetus/embryo develops.
- Explain how traits are passed from parents to their children.
- Identify some problems that can occur during pregnancy.
- Describe the stages of childbirth.
- List the options available for childbirth.

THE CREATION OF NEW LIFE

Human reproduction begins with the act of sexual intercourse. If a woman is ovulating, any sperm deposited inside her body during intercourse may fertilize one of her eggs. A fertilized egg begins the development of a new human being.

Before intercourse itself occurs, both partners usually become sexually excited, often from touching and kissing. Sexual excitement causes the man's penis to become erect, and the inside of the woman's vagina to become moist. The penis is inserted in the vagina. If the man ejaculates, millions of sperm are deposited into the vagina.

How does fertilization occur?

The teaspoonful or so of semen that the man ejaculates usually contains more than 300 million sperm cells. Their long tails move the sperm up the vagina through the cervix, up into the uterus, then into the Fallopian tubes. If an egg is present in one of the Fallopian tubes, the fastest and strongest sperm reach it first, usually somewhere in the upper third of the tube.

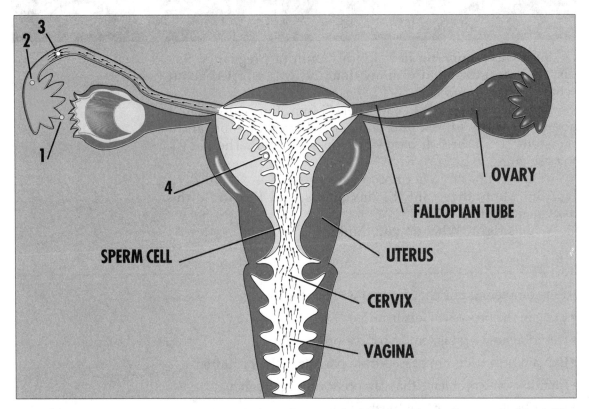

After ovulation (1), the egg moves through the Fallopian tube (2), where it may become fertilized (3). If fertilization occurs, the egg implants itself in the lining of the uterus (4).

Many sperm must reach the egg because it takes their combined effort to burrow through the egg's protective coating. A chemical secreted by each sperm cell helps dissolve the egg's coating. The first sperm to break through the protective coating triggers an instant chemical change in the egg that makes it impossible for any other sperm to enter it.

The sperm that enters the egg releases its genetic material, and the egg is fertilized. This is called the moment of conception (kun-SEP-shun).

The fertilized egg is called a **zygote** (ZI-goht). The zygote begins to grow as it travels down the Fallopian tube towards the uterus. First it divides into two cells, then the two cells divide into four cells, and so on. It takes about three days for the continually dividing zygote to reach the uterus. By the time it gets there, the zygote has become a ball of about 200 cells. The ball of cells is called a **blastocyst** (BLAS-tuh-sist).

Once in the uterus, the blastocyst attaches itself to the lining of the uterus. During the next week, changes in the blastocyst form the **embryo** (EM-bree-oh).

How does an embryo develop into a baby?

At first, the embryo is a hollow ball with two layers of cells. The inside layer will become the baby. The outside layer begins to develop into the **placenta** (pluh-SEN-tuh). The placenta grows a rich supply of blood vessels that lie close to the blood vessels of the uterus.

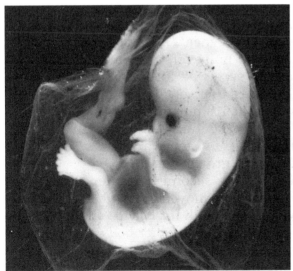 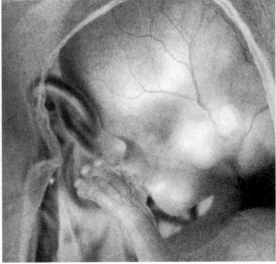

Notice the difference in fetal development from 8 weeks (left) to 16 weeks (right).

The placenta and embryo are attached by a thick rope of blood vessels called the **umbilical** (um-BIL-ih-kul) **cord**. Nutrients and oxygen pass from the mother through the umbilical cord to the embryo. Waste products pass from the embryo through the umbilical cord to the mother. The wastes are disposed of through the mother's lungs and kidneys. A fluid-filled sac called the **amniotic** (am-nee-AHT-ik) **sac** encloses both the placenta and embryo. The fluid acts like a shock absorber, protecting the embryo from jolts.

For the first eight weeks of its development, the embryo is changing rapidly. The major organs, such as the heart, brain, and lungs, are developing. At the end of this eight-week period, all the major organ systems are present, and the arms, legs, fingers, and toes are developed. At this point, the embryo is less than two inches long.

> The normal period of development, from the moment of conception to birth, is called the gestation (jes-TAY-shun) period.

After about 8 weeks after conception, the embryo is called a **fetus**. It is about 3 inches long. The fetus grows rapidly through its remaining time in the uterus. By the end of its normal period of development—38 weeks after conception—the fetus is about 18 inches long and weighs 6 to 7 pounds.

What is pregnancy like for the mother?

During pregnancy, a woman experiences many changes in her body as it adjusts to and supports the growth of the new life inside her. The nine-month period of pregnancy is often divided into three stages called trimesters. Each trimester lasts for three months.

At the beginning of the first trimester, a woman might not even be aware she is pregnant. A missed period may be the first sign. A woman, however, may miss a period for other reasons. The only sure way for her to know if she is pregnant or not is to have a pregnancy test.

During the first trimester, a pregnant woman may experience nausea, fatigue, and tenderness in her breasts. Her breasts also get larger. These changes are caused in part by hormones produced by the growing embryo.

At the beginning of the second trimester, a pregnant woman may begin to feel the movements of her fetus. Soon the fetus has grown large enough for her to "show." During this trimester, the hormone levels in her body stabilize, and usually any feelings of nausea disappear.

During the third trimester, a pregnant woman gains weight rapidly, often about one pound per week. The fetus begins to become large enough to push against the woman's internal organs. As a result, she has to urinate more often and may be short of breath or constipated.

Think and Discuss

1. Where does conception take place?
2. How do nutrients, oxygen, and wastes pass between the mother and embryo/fetus?
3. While the gestation period of humans is 38 weeks, it is shorter for some animals and longer for others. Estimate the length of the gestation period, in weeks, of a cat, a gorilla, a horse, and an elephant. On what did you base your estimates?

WHAT SHOULD YOU KNOW ABOUT GENETICS?

A pregnant woman always wonders what her baby will look like and hopes it will be healthy. These aspects of the baby's physical characteristics are determined mostly at the moment of conception. As you read earlier, at conception, the hereditary information in the sperm cell combines with hereditary information in the egg.

For centuries, people have known that a baby inherits its physical traits from both the father and the mother. But only during the last century have scientists learned a great deal about how the process of genetic inheritance actually works. This knowledge has helped people not only understand such things as how a baby's sex is determined, but also the causes of certain diseases that are passed on from parent to child.

The basis of heredity is a code that exists within every cell of the body. The code is written in the structure of a substance called **DNA**, which exists in units called **genes**. Each gene is a blueprint for the production of a particular chemical substance. By controlling the exact makeup of important substances in the body, genes determine traits such as eye color, height, and the ability to digest the sugar in milk.

DNA stands for deoxyribonucleic (dee-ahk-see-ry-boh-noo-KLAY-ihk) acid.

About 100,000 genes make up a complete set of genetic instructions for a human. These genes are distributed among 23 pairs of structures called **chromosomes** (KROH-muh-sohmz). Every cell in the human body—except eggs or sperm—contains an identical set of 23 chromosome pairs. Eggs and sperm each contain only 23 single chromosomes.

Most genes in the human body are also present in pairs. If the two genes in a pair are the same, they act as one. But if the two are different, one is typically stronger and overshadows the other. In other words, the stronger gene will be expressed as a trait of the individual. The effect of the weaker gene will be hidden. The stronger gene of a pair is called the **dominant gene**. The weaker gene is called the **recessive (ri-SES-iv) gene**.

DOMINANT AND RECESSIVE TRAITS

Genetic Trait	Dominant	Recessive
Hair color	dark hair	light hair
Hair texture	curly hair	straight hair
Eye color	brown eyes	blue eyes
Eye size	large eyes	small eyes
Lip size	broad lips	thin lips
Body height	short	tall
Ear lobes	free ear lobes	attached ear lobes

The gene pair that determines the curliness of hair serves as a good example of how dominant and recessive genes work. If you have two genes for curly hair, your hair will be curly. If you have two genes for straight hair, your hair will be straight. What kind of hair will you have if you have one gene for straight hair and one for curly hair? Since the gene for curly hair is dominant over the gene for straight hair, a person with a mixed set of genes will have curly hair.

As you might have guessed, one gene in each pair comes from the mother, and the other comes from the father. If a mother has genes for both curly hair and straight hair, half of the eggs she produces will have the curly-hair gene, and half will have the straight-hair gene. The genes in the father's sperm cells are distributed the same way. When an egg and sperm come together, each contributes one gene to each gene pair, creating a zygote with a complete set—and a unique combination—of genes.

How are genetic disorders inherited?

Some genes that exist within the human population are defective. Scientists now know that many human diseases are caused by the inheritance of such defective genes. These diseases, called genetic disorders, include PKU, cystic fibrosis (SIS-tik fy-BROH-sis), sickle-cell anemia (uh-NEE-mee-uh), and Tay-Sachs (tay-SAKS) disease.

Fortunately, most genetic disorders are caused by recessive genes. Only a person with two copies of the defective gene will actually have the disease. A person with one copy of the gene won't have the disorder, but can pass the gene on to any of his or her children. Such a person is called a carrier of the gene.

Most defective genes are not very common in the human population. Every once in awhile, however, two carriers of the same defective gene have children together. The chance of any one of their children inheriting two copies of the gene is one in four.

A few genetic disorders are caused by dominant genes. Most of these disorders strike their victims after mid-life. A person with only one copy of the gene will get the disease, but only after he or she has had children, possibly passing the gene on to them.

When two people have family histories of a certain genetic disorder, such as sickle-cell anemia, they should consider getting genetic counseling before having children. With information from tests and family trees, a genetic counselor can determine the chance a couple has of having a child with the genetic disorder. A genetic counselor can also provide the couple with information on whether or not the disease can be treated.

> PKU stands for phenylketonuria (fee-nuhl-kee-tohn-YUR-ee-uh). PKU can cause mental retardation if it is not treated.
>
> Cystic fibrosis: genetic disorder in which the lungs and pancreas don't function properly.
>
> Sickle-cell anemia: genetic disorder in which the red blood cells become abnormally shaped due to a lack of oxygen. Sickle-cell anemia is common among people of African decent.
>
> Tay-Sachs disease: genetic disorder caused by an enzyme deficiency that causes mental retardation, paralysis, and death in early childhood. Tay-Sachs is common among people of eastern European Jewish decent.

How can abnormalities in the fetus be discovered?

There are several types of tests that can be used to determine if a fetus has a genetic disorder or other abnormality.

One safe and common prenatal test, called ultrasound, uses sound waves to form a picture of the fetus. A device may be placed on the pregnant woman's abdomen or in her vagina from which sound waves are beamed at the fetus. These sound waves bounce back and form an image on a monitor like a television screen. Looking at the screen, the woman can see her own fetus moving inside her. Doctors frequently use ultrasound early in the pregnancy to check the position of the fetus and make sure it is developing normally.

> The word **prenatal** means before birth.

Another procedure is a **chorionic villi** (kawr-ee-AHN-ik VIL-i) **sampling (CVS)**. This test is used between the ninth and twelfth weeks of pregnancy to determine if the fetus's chromosomes are normal. In this procedure, a doctor locates the exact position of the fetus with ultrasound. Then with a needle inserted in through the mother's abdomen, the doctor removes cells from the placenta that are then tested for abnormalities.

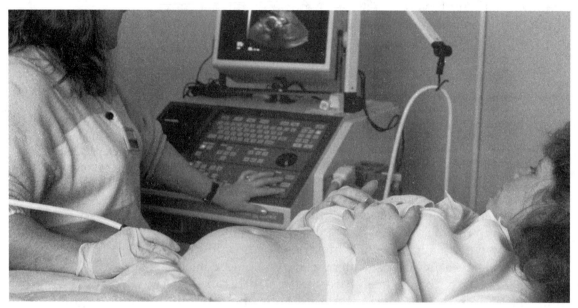

An ultrasound is a common prenatal test to check the developing fetus.

A third type of test is called **amniocentesis** (am-nee-oh-sen-TEE-sis). To do this test, a doctor locates the fetus with ultrasound and then removes a small amount of amniotic fluid through a needle. A laboratory analyzes the cells in the fluid for chromosome abnormalities, or for the presence of certain defective genes. Both chorionic villi sampling and amniocentesis carry risks, so these tests are usually performed only when there is a strong possibility of the fetus having a genetic abnormality.

What determines the sex of a baby?

Sex is determined at conception by the inheritance of particular combinations of chromosomes, not genes. Of the

23 pairs of human chromosomes, one pair is different from the others. The chromosomes in this pair, called the sex chromosomes, exist in two forms, identified as X and Y. Females have two X chromosomes and males have an X chromosome and a Y chromosome.

How do these two combinations occur? When eggs and sperm are formed, only one of the sex chromosomes present in all the other cells of the body goes into each. Thus, every egg a woman produces contains a single X chromosome. The sperm cells of a man, however, contain either an X chromosome or a Y chromosome. If the sperm that fertilizes an egg happens to contain an X chromosome, the zygote will have two X chromosomes and will be female. If the sperm happens to contain a Y chromosome, the zygote will have an X chromosome and a Y chromosome, and will be male.

Think and Discuss

1. What is the difference between a dominant gene and a recessive gene?
2. If a couple finds out through prenatal testing that their child will be born with a genetic disease, what courses of action can they take?
3. Do you think that all pregnant women should undergo prenatal testing? Explain your thinking.

PROBLEMS RELATING TO PREGNANCY

In many ways, pregnancy is as natural as eating and sleeping. All over the world, women go through pregnancy, often with little disruption in their lives, and give birth to healthy babies. Sometimes, however, problems arise in this complicated process of creating new life.

What is prenatal care?

The best way to avoid problems during pregnancy is to receive prenatal care. Prenatal care is a doctor's attention to the health of a pregnant woman and her fetus. Prenatal care is available from clinics, family doctors, and obstetricians (ahb-stuh-TRISH-unz). Obstetricians are doctors who specialize in treating women during pregnancy. During a woman's first visit to a doctor after finding out that she's pregnant, the doctor will evaluate the woman's general health

During pregnancy, a woman should gain between 25 and 35 pounds.

and her body's ability to go through pregnancy. The doctor will also provide advice about diet and other ways to protect the health of the fetus.

Eating a balanced diet is the most important part of a mother caring for herself and her fetus during pregnancy. The embryo and fetus need a constant supply of many kinds of nutrients. Foods that the mother normally eats may not provide adequate nutrition for both her and her developing fetus. Exercise is also important during pregnancy because it helps the mothers' body cope with the stress of pregnancy and reduces extra weight gain.

What substances can harm a developing fetus?

When a woman is pregnant, she must also be especially careful of what substances she puts into her body. Almost anything that gets into her bloodstream will enter the fetus's body, too. Nearly all drugs, including alcohol, the caffeine from coffee, and even aspirin, may harm an embryo or fetus. Before taking *any* drugs during pregnancy, a woman should consult her doctor.

Women who drink alcohol during pregnancy risk serious injury to their babies. Alcohol may damage the developing brain of the fetus. Babies born to women who drank significant amounts of alcohol during pregnancy often have a variety of physical abnormalities and behavioral problems. Since these problems all have a common cause, they are grouped together as **fetal alcohol syndrome (FAS)**.

Addictive drugs such as heroin and cocaine are also dangerous to a growing fetus. The drugs enter the fetus's body and keep it from developing normally. Often, babies born to mothers who use drugs are addicted at birth. They suffer greatly during their first months of life.

Smoking during pregnancy is also harmful to the fetus. The fetus receives less oxygen than normal, as the nicotine enters its body. Women who smoke during pregnancy have babies that, on average, weigh less at birth than the babies of non-smoking women. Many of these low birth-weight babies are unhealthy.

What can go wrong during pregnancy?

There are many problems that can occur during pregnancy over which the mother has little or no control. One such problem is **ectopic** (ek-TAHP-ik) **pregnancy**. This condition results when the zygote attaches itself to the inside of a Fallopian tube instead of to the lining of the uterus. Sometimes an ectopic pregnancy happens because the tube has been blocked by scars from an infection. As the embryo grows, it blocks the blood supply and causes severe pain. Immediate medical care and surgery are needed.

Sometimes a pregnancy ends when the embryo or fetus dies during the early part of pregnancy and is expelled from the uterus. This is called a spontaneous abortion or **miscarriage** (mis-KAR-ij). Miscarriages are often the result of a severe genetic defect in the embryo or fetus. They usually occur during the first trimester. It is common for a woman to miscarry early in a pregnancy and not be aware that a miscarriage has occurred. The woman may not even know that she was pregnant to begin with. A woman who miscarries can usually have a normal pregnancy in the future. Sometimes a fetus develops into the third trimester and then is born dead. This is called a **stillbirth**.

During the last two months of pregnancy, some women develop a serious condition called **preeclampsia** (pree-e-KLAMP-see-uh). The first symptoms are rapid weight gain, swelling, and increased blood pressure. The woman must often be hospitalized so that her diet and medication can be controlled. If she is not monitored, she could have convulsions, go into a coma, or even die. The people most at risk for developing preeclampsia are teenagers and women who have high blood pressure.

Another problem that can occur during pregnancy is **gestational diabetes** (jes-TAY-shun-ul di-uh-BEET-is). When the body is functioning normally, it produces insulin to control the level of sugar in the blood. During pregnancy, some women cannot produce enough insulin to deal with the increased amount of sugar in their blood. The condition is easily treated by controlling the woman's sugar intake. Without treatment, however, there is a 1-in-3 chance the fetus will die. Because gestational diabetes is so common, doctors regularly test for it between the 24th and 28th weeks of pregnancy.

Pregnancy can also be cut short by the early birth of the baby. When a baby is born before the 36th week of pregnancy, its birth is said to be **premature**. The lungs of a premature baby are not yet fully developed and can't function properly. Premature babies usually spend some time in an incubator, where they are kept warm and can breathe oxygen-rich air.

What can be done about infertility?

Some couples have a different type of problem—they cannot start a pregnancy in the first place. This problem is called **infertility**. If two people have tried to conceive a baby for more than one year without success, they may have an infertility problem.

Infertility can have a variety of causes. Many infertile people had a sexually transmitted disease that caused scarring or other damage to their reproductive systems. In older couples, aging is the main cause of infertility. For some couples, the cause of their infertility is unknown.

Either the man or the woman may be infertile. Since doctors must know which person is infertile, infertile couples must first undergo tests to see where the problem lies.

During in-vitro fertilization, an egg is fertilized in a laboratory dish by sperm cells. The embryo then grows in a woman's uterus.

One method used when the man is infertile is **artificial insemination** (in-sem-uh-NAY-shun). In this procedure, the semen of another man, usually an anonymous donor, is placed inside the woman's vagina near the cervix. For fertilization to occur, timing is important: artificial insemination must occur around the time of ovulation.

A method used when the woman is infertile is **in-vitro** (VEE-troh) **fertilization**. Mature eggs are taken from the woman's ovary and placed in a laboratory dish with nutrients and the man's sperm. Fertilization of one or more eggs occurs in the dish. Any fertilized eggs are placed into the woman's uterus, where it is hoped that one egg will implant and develop into an embryo and fetus.

Another way some infertile couples can have a child is through the use of a surrogate (SUR-uh-gayt), or substitute, mother. In this case, the couple makes a legal agreement, or contract, with another woman who agrees to carry the couple's baby in exchange for a payment. That woman is artificially inseminated with the man's sperm. After the baby is born, the surrogate mother releases the baby to the couple.

Surrogate mothering is medically simple, but is legally, morally, and emotionally complicated. There are no hard-and-fast rules governing how it works. In some cases, for example, courts have allowed the surrogate mother to keep the baby after she decided that she couldn't bear the emotional pain of giving it up.

Think and Discuss

1. Why are pregnant women advised not to drink alcohol?
2. What is a miscarriage? What is usually its cause?
3. Sylvia is infertile. Since she can both produce fertile eggs and carry a fetus in her uterus, she and her husband José are considering in-vitro fertilization as a way to have children. What do you think may be wrong with Sylvia's reproductive system?

CHILDBIRTH

Birth is a special event for the mother and father and, of course, for the baby. For the mother, it is also very stressful, painful, and possibly dangerous. Before there were modern medical technologies, many women died in childbirth. For these reasons, choices about how the birth will occur are very important to most expectant couples.

What are the choices for childbirth?

Babies can be born at home or in a hospital. In many countries, including the United States, most couples choose to give birth in a hospital. A small, but growing, number choose to give birth at home. Before the twentieth century, babies were almost always born at home.

Women giving birth have the option of whether or not to receive an anesthetic (an-us-THET-ik), a drug that reduces sensitivity to pain. They may choose to have an epidural (ep-uh-DUR-ul) anesthetic, which numbs the body below the waist. An epidural anesthetic allows the mother to stay awake during the birth, and it usually has no effect on the baby.

Childbirth without an anesthetic is called natural childbirth. During pregnancy, women choosing natural childbirth learn ways of breathing and other methods that reduce the pain of giving birth. They may also learn and practice exercises that prepare their bodies for the stretching and muscular exertion of childbirth.

What happens during childbirth?

After nine months of development, a baby is ready to be born. The process by which the baby enters the world happens in three stages.

In the first stage, muscle contractions in the uterus, each 30 to 90 seconds long, soften and stretch open the cervix. The work performed by these contractions is called **labor**. The process of opening the cervix is called **dilation** (di-LAY-shun). Dilation usually takes at least 4 hours and may sometimes take more than 24 hours.

At the beginning of dilation, the muscle contractions of labor are spaced many minutes apart. By the time dilation is complete, the contractions come very close together, and they are much stronger and more painful. Sometime during this stage, the amniotic sac breaks open and its fluid is released.

The second stage of childbirth begins when the cervix is fully dilated and the baby starts being pushed out of the uterus. During this stage, labor continues and the woman feels an urge to help "push" the baby out with her abdominal muscles. This stage lasts between a half hour and four hours. It ends with delivery, the exit of the baby from the vagina.

Usually a baby is delivered head first. Sometimes, however, the hands, feet, or buttocks emerge first. When this happens, the birth is said to be breech. A breech birth may mean a longer and more painful labor. It may result in harm to the baby if, after the body is delivered, the head is trapped, resulting in the baby being unable to breathe.

Sometimes a baby is delivered through an incision made in the woman's abdomen. This is called a **Cesarean** (si-ZER-ee-un) **section,** or C-section. A doctor may choose to perform a C-section when the baby is having trouble moving through the birth canal, or when the baby is in distress and its life is endangered. Other times, a Cesarean section is planned beforehand as the method of delivery, either by choice of the mother or because the doctor thinks the baby is in the breech position. A C-section is also performed if the mother has a condition such as herpes that can be passed on to the baby in the passage through the birth canal.

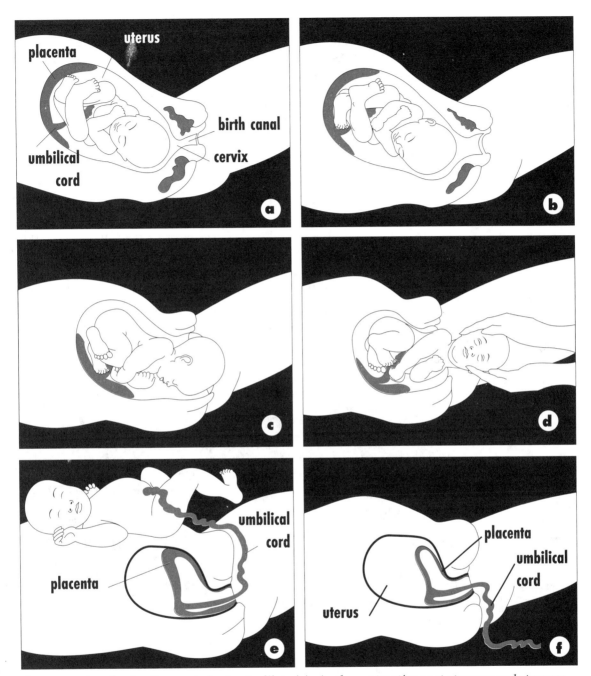

As contractions begin, the cervix begins to dilate (a). As the contractions get stronger and stronger, the cervix continues to open (b), and the muscles begin to push the baby out (c). The next stage of birth is delivery (d) and (e). The umbilical cord is still attached to the baby, as well as the placenta. The final stage of childbirth is the delivery of the placenta (f).

The third and final stage of childbirth is the delivery of the placenta. This organ, which has nourished the baby for nine months, is expelled by labor contractions. It is often called the afterbirth.

What happens after childbirth?

After giving birth, the mother's body goes through many changes. The spongy lining of the uterus breaks down and leaves the body as a discharge. This discharge can last up to six weeks. The breasts begin to produce milk. Because hormone levels change drastically, the woman may feel tired and depressed. The period of time between delivery of a baby and a woman's first menstruation that follows is called the postpartum (pohst-PAHR-tum) period.

Not all mothers choose to breastfeed their infants. There are commercially made formulas that can be purchased that provide the infant with the required nutrients.

The milk produced by the breasts is the ideal food for the newborn baby. It contains the right mixture of nutrients for proper growth. It also contains antibodies, substances that give the baby immunity to some diseases and illnesses. The breasts continue to produce milk as long as the baby nurses.

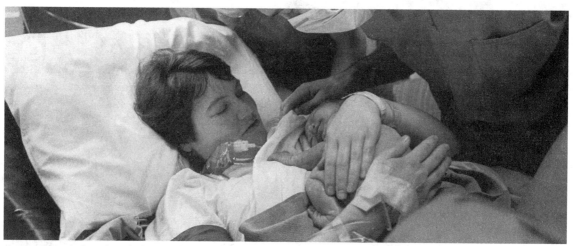

A newborn infant bonds with its parents.

Think and Discuss

1. Describe what happens during each stage of childbirth.
2. What is a Cesarean birth? For what reasons may a baby be delivered this way?
3. A general anesthetic affects the entire body and causes unconsciousness. Why do you think general anesthetics are now only used in emergency situations?

Chapter Summary

- Fertilization takes place when a sperm penetrates an egg. The fertilized egg develops into an embryo and then into a fetus.
- The placenta nourishes the developing embryo/fetus by exchanging nutrients, oxygen, and waste products with the mother's blood.
- Genes, which determine physical traits, are passed from parents to their children through eggs and sperm.
- Several problems may occur during pregnancy, including ectopic pregnancy, gestational diabetes, preeclampsia, and miscarriage.
- Childbirth happens in three stages: the cervix dilates in the first stage, delivery occurs in the second, and expulsion of the placenta occurs in the third.
- Childbirth can take place in a hospital or at home; the mother can receive an anesthetic for the delivery or not.

Activities

1. Find out some details about your own birth. If possible, interview your mother, or other family member. Prepare by writing down at least five questions. After the interview, write a one-page description of your birth from the point of view of yourself as the fetus and newborn infant.
2. Research the childbirth practices in a culture other than your own. How are the fathers involved? Who helps deliver the baby? Is this person a man or woman? Where do births usually occur? Are any ceremonies performed either before or after delivery? What is done to the newborn right after delivery? Present an oral report of your findings.
3. If your local hospital has a birthing center, arrange to visit it. Interview staff nurses or doctors to find out more about how the center operates. Once you collect enough facts, write a fictional story based on fact describing what having a baby at the birthing center would be like.
4. Work in groups to make posters showing stages of fetal development. Each group can be responsible for a poster showing a different stage. Each poster should describe the average size and weight of the embryo or fetus and what is happening at that stage of development. Check a variety of sources, including library books and encyclopedias, before you start making the poster.

5. Set up a debate situation between one group that believes that giving birth in a hospital is preferable to giving birth at home and the other group that believes that home births are preferable to hospital births.

Thinking It Over

1. Daniel and his wife, Katherine, have decided to have children. Just before they stop using birth control, they find out that Daniel's father has developed Huntington's disease, a genetic disorder caused by a dominant gene. This means that Daniel has a 50-50 chance of having the gene himself. Since there isn't yet a test for the presence of the gene, Daniel doesn't know for sure if he has it or not. The possibility scares him and also makes him wonder about having children. A genetic counselor tells them that if Daniel has the defective gene, any child he and his wife have will have a 50-50 chance of inheriting it. A child who inherits the gene will have the symptoms of the disease by the time he or she is 45 and will live only a few more years after that.

Make two columns on a sheet of paper. In one column, list all the reasons you can think of for Daniel and Katherine to have children. In the other column, list reasons why they should not have children. Evaluate both sides of the problem and decide what you would say or do if you were Daniel or Katherine. Explain your decision.

2. Imagine that you're at a family party. You overhear a cousin you admire, in the first trimester of pregnancy, refusing an alcoholic drink. Several other relatives are pressing her to take the drink. They are saying things like, "Oh, come on, take the drink, it'll relax you!" What do you think should be your cousin's response? Write out the dialogue, or conversation, that might occur if this were the situation.

CHAPTER 5
PREVENTING PREGNANCY

Frank and Jed are talking after school. Frank tells Jed that his girlfriend Jenny has been pressuring him for sex.

"What's so bad about that?" Jed asked.

"For one thing, I don't know if I am ready for sex," Frank says. "Plus, I don't know what kind of birth control to use."

Jed replies, "With Tricia and me, I leave that up to her. She knows if she gets pregnant, it's her problem!"

"I'm not leaving it up to anyone else! I figure if I get someone pregnant, it's my problem," Frank answers.

Do either of these guys sound like anyone you know?

OBJECTIVES

After completing this chapter, you will be able to:

- Explain the importance of family planning.
- Describe the various barrier methods of birth control.
- Describe chemical methods of birth control, intrauterine devices, and permanent methods of birth control.
- Describe natural methods of birth control.
- Describe some birth control methods that are being developed for the future.

FAMILY PLANNING IS LIFE PLANNING

When should you have a child? How many children should you have? These are some of the important life decisions a couple must make together. Making these decisions is known as family planning. In order to make family planning work, people use methods of preventing pregnancy, called birth control or **contraception** (kahn-truh-SEP-shun).

contraception: the use of any of a number of methods to prevent pregnancy.

Most sexually active young men and women know about birth control. However, many either do not use any kind of birth control, or fail to use a method properly. This results in many young women having unplanned pregnancies.

Why do some people fail to use birth control?

People fail to use birth control for many reasons. Some people think that a woman cannot get pregnant the first time she has sexual intercourse. Some people fail to use birth control because of misunderstandings about when a woman can become pregnant. They may falsely assume that if the woman has her period, she can't get pregnant. Other false assumptions are that a woman cannot get pregnant right after having a baby or in the months during which a baby is being breastfed.

Some men unfairly think that birth control is the woman's responsibility. Birth control should be the responsibility of both partners. By discussing and sharing this responsibility, partners show respect and caring for each other. Sometimes young people think that wanting to talk about birth control is a signal that a person wants sex or is "easy." The truth is, being knowledgeable about birth control means that a person wants to be able to responsibly avoid pregnancy. Finally, some people think that the use of any kind of birth control is against their religious beliefs. However, there are natural birth control methods that most religions approve of.

Who should use birth control?

If you are sexually active, you should always use some form of birth control. Besides protecting against pregnancy, many forms of birth control give some protection against **sexually transmitted diseases (STDs)**, such as herpes, chlamydia, and HIV.

For more information on sexually transmitted diseases, see chart on pp. 100–101.

Safe, effective birth control is available to anyone who wants it in the United States. Anyone, even a teenager, can obtain some kind of birth control. There are no laws against buying or using contraceptives in most states. Although some contraceptives require a doctor's prescription, other inexpensive contraceptives are readily available in drugstores, supermarkets, and from vending machines.

What is meant by abstinence?

The most reliable form of contraception is **abstinence** (AB-stuh-nuns). Abstinence means not having sexual intercourse. Abstinence is the only method of birth control that is always 100% effective in preventing both pregnancy and STDs. In addition, abstinence doesn't cost anything and is always available.

There are many advantages of abstinence for young people. Abstinence gives you more freedom to plan your future. It assures you of being able to spend time with friends and family. It allows you to love and care for someone of the opposite sex without the pressures created by a sexual relationship. By choosing abstinence, you can complete your education without worrying about pregnancy and STDs. Choosing to abstain does not mean you will be less of a man or less of a woman. Abstinence is a responsible choice.

Think and Discuss

1. What is family planning?
2. What is meant by *abstinence*?
3. Why do you think sexually active people of your age sometimes fail to use any method of birth control?

BARRIER METHODS OF BIRTH CONTROL

Barrier methods of contraception work by blocking the pathway that sperm must take to reach an egg. In order to prevent pregnancy, a barrier method must be used correctly every time sexual intercourse occurs. Some barrier methods require that both sex partners cooperate in their use. If used correctly, barrier methods are very effective in preventing pregnancy. Some are also helpful in preventing the spread of sexually transmitted diseases.

What are condoms?

Male condoms, sometimes called "rubbers," are sheaths made of thin latex rubber or lambskin. A condom works by completely covering the erect penis and collecting the fluid that is released before, during, and after ejaculation. A **spermicide** (SPUR-muh-syd) used with a condom increases its effectiveness.

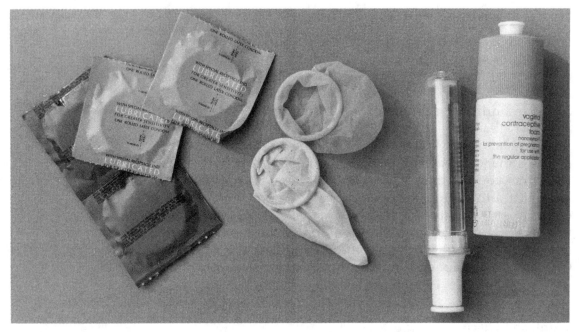

When used properly, condoms offer protection against pregnancy and STD infection.

Lambskin condoms are thinner than latex condoms. They do not provide protection against STDs.

Many people choose to use condoms because they are easy to buy and are quite effective in preventing pregnancy. Condoms made from latex provide effective protection against STDs, including HIV.

To be effective, a condom must be fully intact, with no tears or breaks. It should never be used more than once. To avoid breakage, condoms should be stored in a cool place, away from heat. For example, a condom carried in a wallet, where it is exposed to body heat, should not be used. Heat weakens the rubber and can cause the condom to break or tear. Petroleum jelly also weakens rubber. If a lubricant is needed, only a water-based lubricating jelly should be used.

How do you put a condom on? Open the package carefully so as not to rip the condom. If the condom does not come with a spermicide, squeeze a small amount into the condom. Put the condom against the tip of the erect penis. Make certain to leave about a half–inch of space at the tip where the sperm will be collected. Unroll the condom down to the base of the penis. Keep the condom on during intercourse. Be careful not to tear it with sharp fingernails, rings, or braces.

How should a condom be taken off? After ejaculation, and while the penis is still erect, withdraw the penis from the vagina. As the penis is withdrawn, make sure you or your partner holds the base of the condom so it does not come off. Starting at the base, carefully remove the condom, wrap it in a tissue, and throw it in the garbage. Remember, never use a condom more than once.

The female condom

The **female condom**, or vaginal pouch, is a sheath that covers the vaginal canal, along with the entire genital area. The pouch is closed at one end. It is inserted into the vagina and placed up against the cervix to keep sperm from traveling up into the uterus. The female condom is a preferred means of birth control for some women because it provides more protection against STDs than any other method available for women's use.

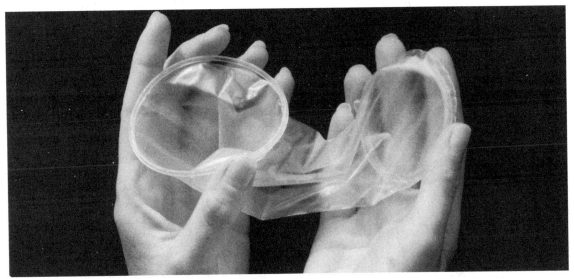

Female condoms have been available since 1993.

Diaphragm

A **diaphragm** (DI-uh-fram) is a dome-shaped cup made of thin rubber. Before use, spermicidal jelly should be placed around the edge of the diaphragm. It should then be inserted by the woman into the vagina and placed against the opening of the uterus. It can be inserted up to 6 hours before sexual intercourse and should be kept in place for at least 6 hours afterwards. A diaphragm should be checked periodically to be sure it has no holes. A diaphragm must be fitted by a doctor or by a nurse-practitioner at a clinic.

Contraceptive sponges

A **contraceptive sponge** is a dome-shaped sponge that contains a spermicide. The sponge is first wet with water, then placed deep inside the vagina covering the opening to the uterus. The sponge must be in place before intercourse begins and must remain in place for at least 6 hours afterwards. It remains effective for up to 24 hours. No prescription is needed for the purchase of a contraceptive sponge.

Cervical cap

The **cervical cap** is a firm rubber cap that fits over the cervix. It must be used with a spermicide and must be fitted by a doctor or nurse-practitioner. It can be worn for up to 48 hours and must be worn for 8 to 12 hours after intercourse.

Spermicides, usually in the form of jellies or foams, contain sperm-killing chemicals. One such spermicide is nonoxynol-9. Spermicides can be used with a condom, diaphragm, cervical cap, or used alone. They work by killing sperm as the sperm enter the vagina. The foam or jelly must be placed in the vagina immediately before intercourse. Spermicides can be purchased without a prescription.

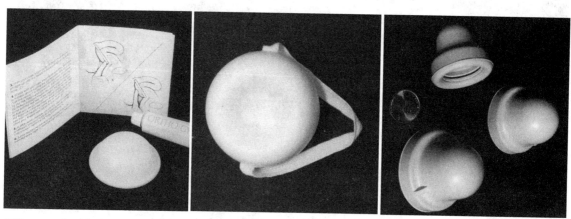

When used with a spermicide, the diaphragm, contraceptive sponge, and cervical cap are more effective.

Think and Discuss

1. Explain how barrier methods of contraception work.
2. Which male and female barrier methods are most effective in preventing STDs? Explain.
3. A friend admits to you that he or she has recently become sexually active and asks you about birth control. What advice would you give?

OTHER METHODS OF BIRTH CONTROL

Nonbarrier methods of birth control include oral contraceptives, the drugs Norplant® and Depo-Provera®, intrauterine (in-truh-YOOT-ur-in) devices, and permanent kinds of birth control. All of these methods of birth control must be obtained by going to a doctor or a clinic.

Oral contraceptives

Oral contraceptives, also called birth control pills, are taken by women. They contain a combination of the female hormones estrogen and progesterone. The hormones in the pills keep a woman from ovulating by tricking the reproductive system into behaving as if the woman is pregnant.

Birth control pills are sometimes prescribed for women who have heavy cramping during their periods, as well as irregular menstrual cycles.

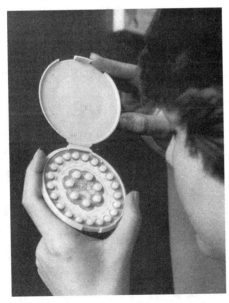

Birth control pills prevent pregnancy by tricking the body into behaving as if the women is pregnant.

63

Usually, a woman using oral contraceptives must take 1 pill each day for 21 days. For the next 7 days, the woman takes sugar pills. During the 7 days with the sugar pills, the woman has her period. Oral contraceptives must be taken each day as directed, regardless of whether or not intercourse occurs. Even 1 missed pill can make it possible for a pregnancy to occur.

Birth control pills are safe for most young women as prescribed by a doctor. However, some women experience mild side effects, such as breast tenderness, nausea, or weight gain. If one kind of pill causes these problems, the doctor can often prescribe another kind. More serious side effects include constant headaches, chest pains, leg pains, or eye problems. Women over the age of 35, especially women who smoke, are at risk for blood clots, which may lead to stroke or heart attack.

Intrauterine device (IUD)

An **intrauterine device (IUD)** is a small piece of copper or plastic inserted into a woman's uterus by a doctor. An IUD prevents sperm from swimming up into the Fallopian tubes and also prevents implantation of a fertilized egg in the uterus. If pregnancy is desired, the IUD must be removed by a doctor.

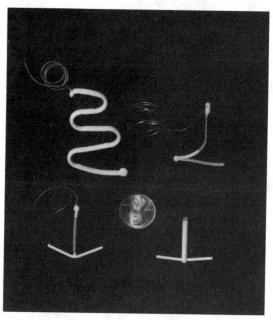

You can see the size of the IUD in comparison to a penny in the above photo.

Norplant®

Norplant® is a set of hormone-releasing capsules that are implanted beneath the skin of a woman's arm through a tiny cut made by a doctor or nurse-practitioner trained in the procedure. The hormones in Norplant® prevent ovulation. Once implanted, Norplant® remains effective for up to five years. The side effects of Norplant® are similar to those of oral contraceptives. If a pregnancy is desired, the capsules can be removed by a doctor.

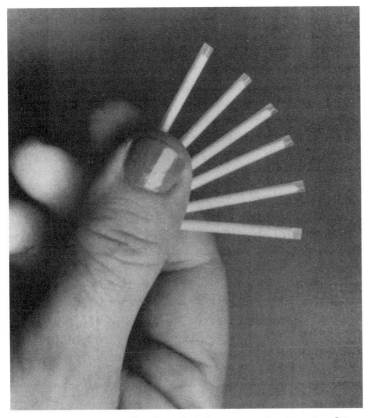

To avoid possible complications, it is important that Norplant® be inserted by someone trained in the procediture.

Depo-Provera®

Depo-Provera® is also a hormone-based contraceptive injection. It is a synthetic version of the female hormone progesterone. Depo-Provera® works by blocking ovulation. Depo-Provera® must be injected every 3 months. The side-effects of Depo-Provera® include irregular periods.

Permanent kinds of birth control

People who are sure that they do not want children or who have completed their families may choose permanent birth control. These methods involve surgery and usually cannot be reversed.

Men can have an operation called a **vasectomy** (vas-EK-tuh-mee). A doctor makes a tiny incision in the scrotum and cuts or ties off the vas deferens, the tubes that carry sperm. After a vasectomy, semen is still produced, but it contains no sperm. A man's sexual desire and his ability to have an erection are not affected in any way.

Women can have an operation called a **tubal ligation** (li-GAY-shun). During this surgery, the Fallopian tubes are tied, or closed off. This prevents eggs from traveling down to the uterus during ovulation, so pregnancy cannot occur. A woman's sexual functioning is not affected by the operation.

Think and Discuss

1. Which birth control methods involve the use of hormones?
2. What is an IUD? How does it prevent pregnancy?
3. What issues do you think a person thinking of permanent birth control should consider before deciding in favor of it?

NATURAL METHODS OF BIRTH CONTROL

Natural methods of birth control involve no devices or chemicals. For that reason, natural birth control methods are acceptable to religious groups who object to the use of other birth control methods. In general, for teenage girls and women with irregular menstrual cycles, natural methods of birth control offer a much lower effective rate than other methods.

The rhythm method

One type of natural birth control is the rhythm method. The rhythm method relies on the fact that women generally ovulate at about the middle of their menstrual cycle. By not having intercourse for several days before and after ovulation, pregnancy may be avoided. However, in practice, this method can be uncertain. Many women do not have precisely regular menstrual cycles, making it difficult to predict the day on which ovulation will occur. To be sure the method will work, the woman must keep a careful calendar of her menstrual cycles, and the couple must abstain from intercourse for about 10 days each month.

The sympto-thermal method

The sympto-thermal method is similar to the rhythm method in that it relies on avoiding intercourse for several days before and after a woman ovulates. During a woman's menstrual cycle, her body temperature varies. When a woman ovulates, her body temperature rises between 0.4°F and 0.8°F. A woman can measure her body temperature daily and track the way it changes throughout her menstrual cycle. Using the record of temperature changes, she can figure out when she is ovulating. Changes in cervical mucus are observed and considered, along with the temperature changes, to predict ovulation. Normally, cervical mucus is a thick yellowish fluid that comes out through the vagina. However, when ovulation occurs, the cervical mucus becomes thinner and clearer.

Sympto-thermal measurements can also be useful for women who are trying to become pregnant.

Unreliable natural methods of birth control

Couples sometimes use **withdrawal** as a birth control method. In withdrawal, the penis is removed from the vagina just before ejaculation occurs. Many teenagers try to use withdrawal because it does not require planning ahead or buying anything. However, withdrawal often fails for two reasons. Generally, a small amount of semen leaks from the penis before ejaculation occurs. Also, many men, especially young men, find it difficult to withdraw in time.

Women sometimes douche (DOOSH) immediately after intercourse to try to prevent pregnancy. Douching involves squirting water, or a prepared douching solution, into the vagina to cleanse it. These efforts are completely ineffective as a method of birth control. Douching after intercourse does not wash out all of the sperm from a woman's reproductive system, and may actually move the sperm farther within it.

What is the future of birth control?

Research is currently being done on ways to improve existing methods of birth control. New designs and new

Method	Estimated effectiveness*	How it works	Availability
Male condom	84%-98%	Prevents sperm from reaching an egg	Over the counter (OTC)
Female condom	75%-95%	Prevents sperm from reaching an egg	OTC
Diaphragm (with spermicide)	82%-94%	Prevents sperm from reaching an egg	Prescription
Sponge	76%-92%	Prevents sperm from reaching an egg	OTC
Cervical cap	82%-94%	Prevents sperm from reaching an egg	Prescription
Spermicides	70%-97%	Prevents sperm from reaching an egg	OTC
Intrauterine devices (IUDs)	96%-over 99%	T-shaped object that contains either copper or progesterone; interferes with sperm transport and egg fertilization	Prescription
Birth control pills	94%-over 99%	Contain hormones that prevent ovulation	Prescription
Norplant®	over 99%	Six thin match-sized sticks placed under the skin of a woman's upper arm. They release a hormone that keeps the body from producing the hormones necessary for ovulation	Prescription
Depo-Provera®	over 99%	Shot that contains synthetic progesterone; blocks ovulation	Prescription
Natural Methods (rhythm, temperature)	53%-91%	Avoiding intercourse around the time of ovulation	Instructions from doctor or clinic
Vasectomy	over 99%	Vas deferens are cut and sealed; prevents sperm from leaving the testes	minor surgery
Tubal ligation	over 99%	Fallopian tubes are blocked or cut, which prevents eggs from reaching the uterus	surgery

* The range in effective rates includes user failure (not using or using incorrectly) as well as method failure (fault with method)

Pros	Cons
easy to buy; side effects are uncommon; offers protection against STDs, including AIDS	need to plan ahead and have product available whenever you have sex; condoms sometimes break; irritation and allergic reaction (rare)
easy to buy; side effects are uncommon; offers protection against STDs, including AIDS	need to plan ahead and have product available whenever you have sex; condoms sometimes break; may be difficult to use; irritation and allergic reactions (rare)
side effects are rare; can be left in place for 24 hours; offers some protection against STDs when used with spermicde	need to plan ahead and have product available whenever you have sex; irritation and allergic reactions (rare); toxic shock syndrome (very rare)
easy to buy; side effects are rare; disposable; can be left in place for 24 hours; offers some protection against STDs when used with spermicde	need to plan ahead and have product available whenever you have sex; irritation and allergic reactions (rare); toxic shock syndrome (very rare); may be difficult to remove
side effects are rare; stays in place for 48 hours; offers some protection against STDs when used with spermicde	need to plan ahead and have product available whenever you have sex; may be difficult to insert
easy to buy; side effects are rare; offers some protection against STDs, including AIDS	need to plan ahead and have product available whenever you have sex; irritation and allergic reactions (rare); messy
very effective; convenient	no protection against STDs; insertion can be painful; may increase cramps and bleeding during monthly period; may become imbedded in uterus; may cause pelvic inflammatory disease (PID) and infertility; perforation of the uterus (rare)
long-term safety; more regular monthly periods; less blood flow and cramps with monthly periods; helps protect against cancer of the ovaries and endometrium; descreases risk of fibrocystic breast disease, ovarian cysts, and PID	side effects may include: nausea, breast tenderness, fluid retention, blood clots, spotting, weight gain, and headaches; some women cannot use (those over 35 who smoke; those with certain medical conditions); offers no protection against STDs
highly effective; convenient; helps protect women from cancer of the endometrium	inserting them or taking them out requires a small cut in the skin; irregular monthly periods; spotting may occur; side effects include: headaches, weight gain; provides no protection against STDs; some women cannot use (those with liver disease, inflammation of the veins, or with a history of breast cancer)
highly effective; convenient; helps protect women from cancer of the endometrium	requires getting a shot every 3 months; takes 6-12 months to regain fertility after stoppage of use; irregular monthly periods and spotting; side effects include: bloating and weight gain, headaches, fatigue; offers no protection against STDs
free	requires frequent monitoring of body temperature and periods of abstinence; hard to pinpoint ovulation
one-time procedure; can be performed in a doctor's office	mostly irreversible; possible pain, infection
one-time procedure	mostly irreversible; possible surgical complications; pain and discomfort; hospital stay is often required

materials may make male and female condoms more comfortable and more effective. Condoms and other barrier devices are being developed that release spermicides. New formulations of contraceptive injections are being developed to have fewer side effects. Efforts are being made to develop contraceptive foams and jellies that are effective against HIV.

Research is also being done to develop totally new methods of birth control. These methods include a birth control pill for men, as well as shots and implants designed to suppress the production of sperm. Basic research is also being done toward the development of a contraceptive vaccine. The vaccine would stimulate a woman's immune system to attack and destroy sperm for a period of months or years. It is anticipated that the male birth control pill and contraceptive vaccine will become available in the next 10-20 years.

Think and Discuss

1. How are the rhythm method and the sympto-thermal method similar? How are they different?
2. Why is withdrawal often ineffective as a means of birth control?
3. Imagine that a drug company has announced the availability of an effective male birth control pill. How do you think such a pill will be received by the public?

Chapter Summary

- Family planning is deciding when to have a child and how many children to have.
- Abstinence is the only 100% sure way to avoid an unplanned pregnancy or sexually transmitted diseases.
- Barrier methods of birth control, such as condoms, diaphragms, sponges, and cervical caps, block the path sperm take to reach an egg.
- Other methods of birth control include chemical methods, such as oral contraceptives, Norplant®, Depo-Provera®, IUDs, and permanent kinds of birth control.
- The rhythm and sympto-thermal methods are natural birth control methods that can be effective if properly used.
- Research is being done to improve existing birth control methods and develop new ones.

Activities

1. Prepare a bulletin board designed to answer the following questions: What are some ways a young person can learn more about birth control? Where can a young person who has questions about sex and birth control go to get accurate information and advice? Write or visit Planned Parenthood or a community health clinic and try to obtain any available booklets, posters, or other written materials that can be added to your bulletin board.
2. Work with three other classmates to brainstorm a list of reasons for choosing abstinence over becoming sexually active. Remember not to criticize each other's ideas. Have one group member read your list to the rest of the class.
3. Some communities allow condoms to be distributed free to high school students at school-based clinics. Plan a class debate on whether or not schools should distribute condoms. One group of students should represent the possible benefits of condom distributions in schools. Another group should represent the possible problems such a program could cause.
4. The drug RU-486 is currently being used in Europe to terminate unwanted pregnancies in the early weeks of pregnancy. At the present time, it is undergoing clinical trials in the United States. People opposed to abortion rights do not want RU-486 approved by the FDA. However, the drug has potential medical uses for birth control, as well as for the treatment and prevention of breast

and ovarian cancers. Women's health advocates, scientists, and many physicians think that RU-486 should be approved by the FDA. Hold a class debate on the question: should RU-486 be approved by the FDA?

5. Work with two or three other students in a small group to design a poster comparing barrier methods of birth control, nonbarrier methods, and natural methods.

Thinking It Over

1. Below is a list of statements that could be used to try to persuade someone into risking an unplanned pregnancy. Copy the list of statements into your notebook. For each statement, write how you would resist the pressure.

Statements

   a. Everybody we know has already done it. Why should we wait?

   b. You can't get pregnant the first time we do it.

   c. It won't be a problem; we'll use withdrawal.

   d. Condoms spoil the feeling.

   e. It's OK, I took one of my friend's birth control pills.

   f. We did it before and nothing happened.

2. Teenage pregnancy is a big problem throughout the United States. How do you think an unplanned pregnancy would affect your life? How do you think it would affect your family or family's life? How do you think it would affect your relationship with your friends? How might it affect your plans for the future? Write a short essay answering these questions.

3. Imagine that you have been dating the same person for almost a year. You feel strongly that you want to wait to have sex, but your boyfriend or girlfriend is pressuring you. You try to convince him or her that remaining abstinent is the best option for both of you at the present time. Develop a skit acting out this situation. Perform your skit for the class. Afterwards, ask for additional suggestions from that class on how you could persuade your partner to remain abstinent.

CHAPTER 6
THE CHANGING FAMILY

You just can't stand it anymore. You haven't had a minute's privacy since Laurie and her father moved in. You were glad for Mom when she and Ralph announced that they were getting married. But then you realized that Laurie would be coming to live in your house—and that she would be sharing *your* room!

Your friend, Jenny, hears your daily complaints about sharing your room with Laurie. Finally, Jenny says, "Wouldn't it be great if you could come live in my house? My brother is going to college in the fall, and you could move into his room. I'll talk to my parents about it right away."

Would it be great? What do you think?

OBJECTIVES

- After completing this chapter, you will be able to:
- Tell what is meant by a family.
- Describe some different kinds of families.
- Understand why some people marry and why others do not.
- Describe the responsibilities of being a parent.
- Identify some of the problems involved in teenage marriages and pregnancies.

WHY DO WE LIVE IN FAMILIES?

In addition to food, clothing, and shelter, what do all people need? We need to feel loved and protected. We need to be taught and respected. At its best, a family provides its members with all of this. Our families also help us to develop a sense of self—self-confidence, self-determination, and self-esteem. Self-esteem is the feeling of being a capable and worthwhile person.

A family is made up of all the people who are related, either by blood, marriage, or by law. The people with whom we live make up our family.

At one time, people thought of a family as one that is made up of a father, a mother, and the children of these parents. The father went outside the home to work. The mother stayed home to care for the children. While some of today's families are just like this one, statistics show that most are not.

Is there a typical family?

In recent years, there have been many changes in social attitudes, values, and economics. The makeup of the family reflects these changes. More and more, both parents in a family have to work in order to earn enough money to pay for a family's needs.

There are many different kinds of families.

Some people live as part of an **extended family**. An extended family is one in which other family members, such as grandparents, aunts, uncles, or cousins, are a part. This lifestyle sometimes allows for one adult to look after the needs of the family at home, while all the other adults work outside the home.

Whether as a result of divorce or separation, the death of a parent, or for other reasons, many women and men head a **single-parent family**. Not having another adult to share the work, the costs, and the other demands of being a parent can put an extra-heavy burden on the single parent. The children may share this burden. Of course, having a warm, caring relationship with a parent is more important than the number of parents a child has.

When the members of two families come together, as when a man and woman remarry, a **blended family** is formed. This new relationship may mean happiness for some family members, but not for others. Tensions may arise when people who are used to their own ways of doing things are unwilling to make changes. If everyone shares his or her thoughts and feelings and is willing to compromise, the blended family can become very strong and supportive.

Some adults who choose to live together as a family may also choose not to get married. They may be members of the opposite sex, or they may be members of the same sex. In each situation, couples may decide to have children. Sometimes they become parents to the children born to one or the other of them.

However many members a family may have and whatever the gender of its members, as long as love, caring, and mutual trust are basic to its structure, it is a functional family. The members of a functional family feel important and valued for just being themselves.

Think and Discuss

1. What is a family?
2. How does an extended family differ from a blended family?
3. The word *dysfunctional* means not functioning. What kind of behavior might characterize a dysfunctional family?

WHY DO PEOPLE MARRY?

When teenagers overhear their parents having an argument, they often worry that the fighting will lead to separation or divorce. Sometimes it does. More often, however, the disagreements are resolved.

As with any human relationship, there can be a down side to marriage. Most marriages, however, have the potential for being happy, warm, and supportive relationships. Whether most marriages are happy or unhappy, the institution of marriage is clearly here to stay. Many of the adults you know are married. Many of the young people you know assume that they will marry someday.

What is meant by commitment?

A commitment is an ongoing promise to do something. Sometimes one makes a short-term commitment, such as taking on a regular baby-sitting job. Sometimes one makes a longer-lasting commitment, such as a relationship with another person.

This example of a blended family includes children from previous marriages, and children of this marriage.

Some couples decide to marry because they want to live together and have a sexual relationship. Out of respect for the traditions of their family or their faith, they would feel uncomfortable in the relationship without being married. Although this may be a good reason for choosing marriage, it may not be enough to sustain a lasting marriage.

Perhaps the strongest reason why people marry is the desire to raise a family. In general, people believe that marriage is the most desirable situation in which to have children. Again, the couple must be sure that their commitment to each other goes farther than just the wish for a family.

Why do some marriages work better than others?

Recent headlines have claimed that 50% of all marriages end in divorce within a few years. This information, however, is misleading. It is true that about half of all *new* marriages end in divorce. The divorce rate drops to about 12%, however, when the many previously entered-into marriages are taken into account.

Either way you look at it, a great number of marriages do not work out. It might be that the individuals involved in some marriages are less committed to each other. It could also be that the individuals involved are less able to make their marriages work.

So what do adults do to make their marriages work well? People who have long-term, successful marriages usually are good at compromising. They communicate well with one another. They trust each other. In addition, they work at making their marriage a success.

Why do some people stay single?

There are many reasons why people remain single. Some may pursue demanding jobs and careers. Others may decide that they are happy pursuing dreams and goals alone. Still others may have had childhood experiences that have made them feel that marriage is too risky to pursue.

Whether a person chooses never to marry, or chooses to remain single after having been married before, one thing is certain—it is okay for an adult not to be married.

As you know, some people who are unmarried have children—either children whom they have borne themselves or children whom they have adopted. Whether the unmarried adult heads a family or not, he or she can live a happy and successful life without being married.

Think and Discuss

1. What are some good reasons for a couple to marry?
2. How can a married couple help each other grow?
3. What do you think might account for the higher rate of new marriages ending in divorce?

BEING A PARENT

You have learned how easy it is to *become* a parent, but do you know how hard it is to *be* a parent? It may seem that being a parent means having some freedoms you wish you had right now. For example, as a parent, you may have the freedom to do what you like and to go wherever you like. Being a parent, however, has a lot less to do with enjoying freedoms than it does with taking on responsibilities.

People who want to be parents should be emotionally mature. Their personal relationships should be loving, secure, and stable. They should have a good sense of what they want to accomplish for themselves in terms of education and employment. Dealing with a new baby involves not only day-to-day care, but also long-term physical, social, and emotional care. For this reason, only those people with the maturity to take charge of their own lives can give a child everything he or she deserves.

Deciding whether or not to have a child is one of the most important decisions most couples will ever make. They should ask themselves questions such as: How will we raise a child? Are we willing to change our lives for a child? Are we financially able to support a child? If there is uncertainty in the answers to these questions, a couple should recognize that they may not be ready to have a child.

What do children need from their parents?

Children need parents they can learn from, depend on, and model their behavior after. When children see a caring and loving relationship between two people, it teaches them how to establish strong bonds between themselves and others. The way feelings are treated within the family affects how children feel about themselves.

Things to consider before becoming a parent

For some adults who are totally preoccupied with their professions, their hobbies, or their free-and-easy social lives, a child may come as an intrusion. Child-rearing takes a great deal of a parent's time. It is also costly. Raising 1 child from birth to age 18 costs about $100,000. Adults who are financially unwilling or unable to afford a child might make the responsible choice not to have one.

> **CHILDREN ARE ENTITLED TO...**
> - **A secure and stable environment.**
> - **A caring and loving relationship.**
> - **Consistent responses and feedback.**
> - **Personal time and attention.**
> - **Clearly established and enforced rules.**
> - **Open lines of communication.**

Some couples who are having problems may think that having a child will bring them together. After the baby is born, however, they may find that the burdens of parenthood have put new pressures on them and have caused their problems to worsen.

Unless children are a priority in a couple's plans, they may be wise to decide not to have children. Instead, they may enjoy spending time with the children of relatives and friends. This way, they can have the pleasure of being with a child, without assuming responsibility for the child's well-being.

Think and Discuss

1. What should a person know about him or herself before deciding to become a parent?
2. How can the lack of consistent responses and feedback from a parent affect a child?
3. Describe the characteristics of a couple who you think make good parents. (It may be either a real couple or one of your imagination.)

When teenagers fall in love, whether or not they are having a sexual relationship, one or both members of the pair may want to get married.

You have just learned that the divorce rate for all new marriages is very high. The failure rate of teenage marriages is even higher—over 70%! Building a good marriage is difficult for adults, but it is even more difficult for teenagers.

Early marriage may mean dropping out of school and going to work in order to support a family. For financial reasons, married teenagers may need to live with the parents of one of them. Such an arrangement can be a problem for many reasons. One such problem is that both the teen couple and the parents may feel that they do not have the privacy they need.

Many teenage marriages don't work because the partners feel that by getting married they had to give up their freedom and their friends.

So why do teenagers marry?

One reason for teens wanting to marry is to gain independence from their families. If they feel unloved by their parents, or are unhappy or insecure in their families for other reasons, they may act impulsively. They may see marriage as an acceptable way to run away from their problems. This never works. People can never run away from their problems. Teenagers who go into a marriage expecting an end to their problems are very likely to be disappointed.

Of course, some teen marriages do work out. Like other marriages, if the partners are considerate of one another, if they communicate and share the responsibilities of the marriage, their marriage can work.

What if you're pregnant?

Some teenagers mistakenly use sex to gain a feeling of closeness they lack in their homes, to express anger, or to be accepted. Some allow themselves to be pressured into having sex. While today's teenagers have learned how sexually active people can prevent pregnancy, they sometimes take an "it can't happen to us" attitude and fail to use contraception.

Most teenage marriages take place after couples find out that they have started a pregnancy. Each year in the United

States, more than 1 million teenage girls become pregnant, and 600,000 babies are born to them. Most of the babies are unwanted. If a couple does not want the pregnancy, marriage is probably not the right course to take.

Being a parent is different for a teenager than for an adult. Teenage parents and their babies have more physical, emotional, and social problems than do adult parents and their babies.

Babies born to teenage mothers are more likely than other babies to be born prematurely and at low birth weights. Many have birth defects and learning disabilities. Some babies of teenage mothers are born with afflictions resulting from their mothers' poor prenatal care, diet, and the use of cigarettes, alcohol, and illegal drugs.

While the odds are against a teenager raising a child, it can be done successfully. Success depends on getting strong support from families and from counselors at school, clinics, or social agencies such as Planned Parenthood.

Teenage mothers have a more difficult time raising their babies.

Think and Discuss

1. Identify three reasons why teen marriages often fail.
2. What are some health problems that can affect a pregnant teenager and her baby?
3. What are some reasons that would keep you from entering into a teenage marriage?

WHAT OTHER OPTIONS BESIDES BEING A PARENT ARE AVAILABLE?

If a pregnant woman is not ready to be a parent, she may choose to place the child up for adoption. **Adoption** is the legal process by which individuals gain the right to raise someone else's child as their own. Adoption agencies are staffed by social workers whose job it is to match couples with children available for adoption.

A birth mother may choose to have an open adoption, which allows her to have some say about the couple with whom her baby is placed. Alternatively, she may choose not to become involved, as with a closed adoption. This leaves the placement decision to an adoption agency, while protecting her privacy. Although a decision to place a child for adoption is usually a painful one, it is often in the best interest of the child who is the result of an unintended pregnancy.

"What if I just don't want to be pregnant anymore?"

The removal of a growing embryo or fetus from the uterus before it fully develops is called an **abortion.** At one time, abortions could not be performed legally in this country. In 1973, however, the Supreme Court decision, *Roe vs. Wade,* allowed for abortions to be performed legally. Even though it is a legal procedure, it is still controversial. Some people believe that aborting a fetus is the same as killing a person. Others disagree and say that whether or not to abort a fetus should be the decision of the pregnant woman. Whatever one's attitude toward abortion, there is general agreement that abortion should never be used as a form of birth control.

An abortion is a medical procedure done to terminate pregnancy, that is performed by a doctor. The method of abortion used depends on the stage of pregnancy at which the abortion is done. Those abortions performed during the first trimester are the simplest and the safest. If it is very early in the pregnancy, within three weeks of a missed period, a menstrual extraction can be done. A small tube is inserted into the uterus through which the contents of the uterus are sucked out. This procedure can be done without anesthesia.

Vacuum aspiration (VAK-yoo-um as-puh-RAY-shun) is another procedure that is used until the end of the third month of pregnancy. It is similar to the menstrual extraction method. Due to the increased size of the embryo, however, the instruments used are somewhat larger. This procedure is done under local anesthesia.

Vacuum aspiration is also known as the suction method.

Abortions can be performed beyond the third month of pregnancy by two different means. A **dilation and curettage** (kyoor-uh-TAHZH), or D and C, is an opening of the cervix (dilation) followed by a scraping of the inside of the uterus (curettage). This must be done under general anesthesia. Side-effects of a D and C may include hemorrhaging and infection.

A D and C has other medical uses, besides abortion. For example, if a pap smear comes back abnormal, a doctor will perform a D and C in order to get a sample of the cells.

It is very important for a pregnant woman who is thinking about having an abortion to have help in making her

The issue of abortion is very controversial.

decision. She should try to have the support of people close to her. Someone considering an abortion should speak to a counselor at an agency, a health clinic, or a hospital to be sure that the decision is right for her. Abortions should only be done in state-certified abortion clinics or hospitals.

Think and Discuss

1. If a birth mother chooses to place her baby for adoption, what two methods of adoption are available to her?
2. Why is the menstrual extraction method of abortion more desirable than other methods?
3. If someone your age confided to you that she thought she was pregnant, what would you advise her to do? Why?

Chapter Summary

- A family is made up of all the people with whom you live and all the people who are related to you, either by blood or by marriage.
- Some families may be extended families. Others may be single-parent families; still others may be blended families.
- When people fall in love, they may choose to enter into a marriage in which they commit themselves to share the responsibilities of adult life. Other people may choose to remain unmarried.
- A responsible parent is one who is emotionally mature and who can give a child a loving, secure, and stable home.
- Most teenage marriages, especially those that take place because a pregnancy has been started, are not successful.

Activities

1. Interview a person in your classroom or in your community whose family is different from your own. Ask questions such as: Who works outside the home? Who does the work around the house? How do the members of the family help each other? Write a one-page report comparing this family to your own.

2. Suppose your mom is yelling at your brother because he told her he just found out that his girlfriend is pregnant. Mom says, "How could you be so careless! Now what do you two plan to do about this?" Matt yells back, "Who said we have to do anything now? We'll just have a baby!" As a class, brainstorm some ways to help them have a less emotional, more meaningful interaction to discuss what the family might do to handle their situation. (Be sure not to take sides.)

3. Work in small groups of three or four. Talk about what your family and your religious group have taught you about how sex is related to love and marriage. Tell which of these teachings is most meaningful to you. Are any of them not meaningful to you? Give reasons for your answers.

4. Do you have any unresolved problems at home? Are you unhappy with the way anything is done, or angry about how something is handled? Perhaps you wish that some family members would not always be "at each other." Write a letter about whatever is bothering you to someone in the family. Then, "send" it—or don't!

5. Make a list of personality characteristics that would be important to you in a marriage partner. Consider factors about the person that involve physical appearance, level of education, sexual demands, and attitudes about emotions, money, cleanliness, and just about anything else you can think of!

Thinking It Over

1. Imagine that this is your own situation. Ever since you were a small child, you have known that you were adopted. Your adoptive parents explained why they wanted to adopt a child. They often tell you how happy they were when you came to live with them and how lucky they feel to have gotten you. Recently, you have been wanting to know what your birth parents were like. Since your adoption was closed, however, your adoptive parents don't know anything about your birth parents. You love your parents very much. You don't want them to feel hurt by your desire to find your birth parents. Work with two other students to role–play the situation. Begin the discussion by telling your parents that you have heard about adoptive search groups and that you would like to contact one. You and the other students should think carefully ahead of time about the sensitivity of your statements and questions. As you progress through the conversation, be sure to show concern about each other's feelings.

2. Assume that you are an abortion counselor in a health clinic. You had three counseling sessions in one day. Pick one of the three situations below to think about. Then make a list of issues that you think those involved should consider in making their decisions. Remember: a good counselor makes information available, but does not offer personal opinions.

Situations

   a. A 17-year-old girl and her husband who is 19.

   b. A 15-year-old girl who doesn't want her parents to know about the pregnancy.

   c. A couple who are both 16. The girl doesn't want to keep the baby. The boy wants to marry the girl. He also wants her to keep the baby.

CHAPTER 7
SEX AND SOCIETY

High school seniors, Becca and Philip, have been a couple since their sophomore year. One night, Philip and Becca decide to take a walk by the lake. They sit on a bench to talk. Soon they are kissing and making out. Philip undoes Becca's bra. Becca doesn't respond. When he reaches under her skirt, though, Becca pushes his hand away.

"You want it just as much as I do!" Philip hisses.

His rage takes them both by surprise.

"Philip, stop, now!" Becca yells.

Philip hesitates. Something inside tells him to stop before it is too late. For a moment, Philip loosens his hold on Becca. If you were Philip, what would you be thinking? If you were Becca, what would you do next?

OBJECTIVES

After completing this chapter, you will be able to:

- Identify and describe some unhealthy sexual behaviors.
- Outline an effective strategy to protect yourself against sexual assault.
- Analyze how advertising and the media influence people's ideas about sex.

WHAT ARE SOME UNACCEPTABLE SEXUAL BEHAVIORS?

Have you ever heard the words, *flasher*, *peeping Tom*, or *hooker*? What do these words mean? Why do they make most people feel uncomfortable?

What is exhibitionism?

An **exhibitionist** (ek-suh-BISH-un-ist) is someone who gets sexual pleasure by "flashing" his or her genitals. Exhibitionists are usually men. When you hear the word *flasher*, you may think of a seedy-looking man in a raincoat. This stereotype, however, is misleading. Exhibitionists may be young, handsome, well-dressed, rich, or poor.

Exhibitionists are looking for the shocked or surprised reactions of their victims. The only way to deal with an exhibitionist's display is to look away, then to be sure to move quickly and calmly away. Walk toward a group of people, preferably toward a crowd. Then, as soon as possible, report the crime to the police.

Exhibitionism is illegal. Unfortunately, many exhibitionists cannot keep themselves from behaving in this way. Even if they are caught, they may later repeat their crimes.

Voyeurism

A peeping Tom, or **voyeur** (vwah-YUHR), gets sexual pleasure from watching other people as they undress or have sexual intercourse. A voyeur may be a stranger, or even a friend or relative who frequently hides near an open bedroom door as you are dressing.

Unlike exhibitionists, voyeurs try to remain hidden. Simple precautions can protect you from the unwanted attention of a voyeur. For example, before getting dressed or undressed, close the door to your room and the curtains or window shades. Look around you before changing clothes in a gym, swimming pool, or other public facility. If you feel that a relative may be a voyeur, make sure you are never alone in the house with this person.

> The expression "peeping Tom" comes from an old English legend. According to the story, Lady Godiva rode through town in the nude in protest of her husband's taxation of the poor. "Tom the Tailor" was the only one in town who peeped through his shutters at Lady Godiva.

Prostitution

People who engage in exhibitionism and voyeurism do so to satisfy their sexual desires. **Prostitution** (prahs-tuh-TOO-shun) differs from these behaviors in part because prostitutes are economically, not sexually, motivated.

> **prostitution:** the exchange of sex for money.

Approximately 1.3 million women in the United States earn their living as prostitutes. Many men also become prostitutes, as do teenagers and children.

Most teenagers enter prostitution as a direct result of running away from home. Because they are desperate for love, security, and money, they become easy victims of procurers, or pimps. Pimps are people who find people for the purpose of prostitution.

As a prostitute, teenagers face numerous dangers, including drug dependency, physical violence, and exposure to sexually transmitted diseases such as AIDS. Instead of being rewarded by his or her job, a prostitute is likely to become a victim of it.

Prostitution is illegal in almost every state in the United States.

Think and Discuss

1. What is meant by *exhibitionism* and *voyeurism*?
2. What are some dangers involved in prostitution?
3. Some people fear that if they report a flasher to the police, the flasher might find them and try to attack them sexually at a later date. Should this keep a victim of a flasher's behavior from reporting him? Explain.

WHAT IS MEANT BY SEXUAL ABUSE?

Child abuse, **incest**, and **rape** occur frequently. It is possible that you know someone who has been, or is being, sexually abused. You may find reading this section to be very upsetting, especially if that person is you.

As you read about various forms of sexual abuse, be aware that the victim is *never* to blame. A child molester (muh-LEST-ur) may tell young children that the violation of their bodies is somehow their fault. A rapist may say that the victim wanted it. *Remember: no one is guilty for another person's attack on his or her body. Blaming the victim only allows sex criminals to get away with their crimes.*

child abuse: physical, sexual, or emotional assault by an older person on a child.

incest: sexual abuse, often involving sexual intercourse, that occurs within a family.

rape: a sexual assault committed by violence, threat, or persuasion that culminates in sexual intercourse.

How are some children abused?

A mother ties her son to the radiator "because he threatened to run away." A 17-year-old camp counselor kisses a 12-year-old camper on the lips. A stepfather photographs his wife's 6-year-old daughter in the nude. In each of these incidents, a child is abused. Most child abuse occurs within the family.

Not all abuse is violent, or even physical. Angry words can be as hurtful as physical abuse. Constant or repeated screaming at a child is emotional abuse. The emotionally abused child usually has poor self-esteem.

Responsible parents teach their children to protect themselves against abuse. Even very young children should be taught to refuse rides from strangers. They should be taught to say "no" to activities that involve keeping secrets about their bodies, and to tell an adult if someone touches them in a sexual way. Some parents even enroll their children in self-defense classes.

Victims of abuse need the courage to tell someone who can make the attacks end. Child abuse, however, often destroys this courage. Victims of abuse often think that they are responsible for the abuse, especially when the abuser is a parent.

If you feel that you are the victim of abuse—physical, sexual, or emotional—try to find one person whom it would be safe to tell. That person might be a brother or sister, a teacher, a member of the clergy, a health-care provider, or an adult friend.

What is rape?

Rape is a violent act in which one person forces another to have sex. Usually the rapist is a man, and the victim a woman. Here are some common myths and facts that people have regarding rape.

Some people think that girls and women who wear short skirts or other "sexy" articles of clothing are inviting rape. In fact, rape is an act of violence, not of sexual desire. What a woman is wearing, or even how attractive she is, has little to do with whether a rapist attacks her. Instead, a rapist chooses his victims by how easily he can overpower them. An elderly woman who appears weak might be an easier target than a teenage girl in a miniskirt whose posture, walk, and manner signify that she is strong.

Some people think that boys cannot be raped. In fact, there are three types of sexual intercourse: vaginal, oral, and rectal. Rape results when any one of these occurs through threat, persuasion, bribery, or force. A rapist can force his penis into a boy's mouth or rectum. Furthermore, boys can be sexually assaulted by women.

Male rape victims usually do not tell anyone about a sexual assault. Some boys fear that they will not be believed. Others fear that people will think they are homosexual. Boys who have been brought up to believe that they should always be ready for sex may be embarrassed to admit to being raped.

Some people think that rape most often occurs outdoors in deserted public places, such as parks. In fact, rape most often occurs indoors in familiar places, such as the homes of the victims.

Some people think that rapists and rape victims are usually of different races. In fact, this fallacy is based on racial prejudice. The truth is that nine out of ten rapists and their victims belong to the same racial group.

Some people think that most people who are raped do not know the person who raped them. In fact, most rapes occur between people who have at least some prior acquaintance. A rapist may assault a neighbor whom he observes over time. A rapist and rape victim may have a professional relationship in which the rapist has authority or power, such as that of a doctor and patient.

PREVENTING SEXUAL ASSAULT AND RAPE

- Never walk alone at night.
- When you walk down a street, look as if you know where you are going and what you are doing.
- Be aware of your surroundings.
- Pay attention to your instincts.
- Avoid deserted streets.
- Wear shoes or sneakers that you can walk or run in easily.
- Don't wear earphones that can block your hearing.
- If you are being harassed, ignore the comments and leave the situation.
- Look in the front seat and the back seat of a car before getting in.
- Carry your keys in your hand when going home or to your car.
- Don't hitchhike or take rides from strangers.
- Whenever possible, tell someone where you are, who you are with, and when to expect you.
- If you think you are being followed:
    - turn around and look.
    - move to a busier area.
    - go into a store or public place and ask for help.
    - yell to attract attention.

In other instances, the relationship is even closer. Rape can occur between a parent and child and between brother and sister. It can also occur between married people and in homosexual relationships.

Many people have difficulty understanding that rape can occur within marriage. A husband who rapes his wife may believe that because they are married, she has no right to refuse to have sex with him. This is untrue. All men and women have the right to determine how and when they are touched and who may touch them. Whatever two people's relationship to each other may be, if forceful sexual intercourse takes place against the wishes of one of them, it is rape.

How can you avoid date rape?

You have the right to say "yes" to some sexual activities and "no" to others. If your date persists after you say "no," it is assault. If your date's persistence results in sexual intercourse, it is rape.

Here are some ways you can try to protect yourself against **date rape**. Begin by choosing your dates carefully. Pay attention to warning signs, such as a date's making fun of your sexual limits, or persisting in going farther when you have said "no." Avoid dating people who suggest that you owe them something because they have spent money on you. Date rape experts believe that both parties should pay their own way on dates. When two people share the costs, they also share the decision-making about how to behave on the date.

Suppose, despite your efforts to choose carefully, you get a creepy feeling after the evening has begun that your date may become violent. Pay attention to your feelings. If something feels wrong, it probably will be wrong. Don't be afraid to leave your date in a movie theater or at a party and call your parents to drive you home. You are much better off being embarrassed than to risk being raped.

Be sure to avoid the kinds of situations in which date rape most often occurs. Do not leave a social event with a person you do not know well. If you agree to go outside or upstairs with someone you meet at a party, tell a friend exactly where you are going and when you expect to be back. Ask the friend to come and get you if you don't show up at the appointed time.

Although drinking doesn't always lead to date rape, date rape almost always involves alcohol. If you are unwilling to avoid alcohol altogether, strictly limit your drinking and make sure that your date does the same. Never get in a car with someone who is high on alcohol or other drugs.

A rape that involves sexual intercourse between an adult and a person who is a minor is called statutory (STACH-oo-tawr-ee) rape. Even if the minor is a willing partner, if he or she is below the "age of consent," the adult could be found guilty of statutory rape if charges are brought against the adult.

The age of consent in most states is 18.

What should you do if a rape occurs?

If you are raped, whether by an acquaintance or a stranger, you need to get immediate medical attention. Do not wash your body, or change your clothes. If you decide to take legal action against the rapist, semen and other evidence of injury can be used as proof of the rape.

If possible, ask for your parents' or guardians' help in getting you through the aftermath of a rape. If they cannot, or will not help you, phone a rape-crisis center.

You can look in the phone book under "Rape" for the local rape-crisis center.

A rape-crisis center will provide, at no cost, the services of a highly trained counselor. The counselor will go with you to the hospital, make sure that you get proper medical attention, and arrange for follow-up support. Counseling is very important to most rape victims' recovery. A counselor can help you to express the powerful feelings of anger, guilt, and fear that accompany rape. Over time, the counselor will help you regain confidence and self-esteem.

What is sexual harassment?

In 1991, law professor Anita Hill accused U.S. Supreme Court nominee Clarence Thomas of **sexual harassment** (huh-RAS-munt). Many people watched the televised Senate Judiciary Committee hearings of Hill's testimony. Many became newly aware of how widespread sexual harassment is and became determined to try to stop it.

sexual harassment: any form of unwanted sexual attention, including leering, pinching, lewd comments, and propositions.

Sexual harassers usually have power or authority over the people they harass. A welfare worker may make sexually suggestive comments about a client's body or clothing. A professor may suggest that a student can improve his or her grade by having a drink after class.

Sometimes people are not sure if certain treatment is sexual harassment. They may wonder, "Is it real, or am I imagining it?" If you feel that you are being sexually harassed, describe the behavior to an adult whom you trust. Check with peers who also have contact with the harasser. Ask if they have observed or experienced the same behavior. Keep a log of harassing behaviors. Record each incident and the time and place of each. Write down the names of any witnesses to the behavior. Tell the harasser, calmly, clearly, and directly, that you are not interested in his or her attentions. Finally, don't blame yourself. You did not cause the behavior. The sexual harasser is the one who should be embarrassed and who must be held responsible.

Sex and technology

Some sexual harassers, instead of facing their victims, confront them on the telephone. They make an obscene phone call by dialing a number, then try to engage whoever answers the phone in a conversation about sex. Some obscene phone calls consist of the caller's making sexually explicit remarks into the phone without waiting for a response. Some callers breathe heavily into the phone without speaking at all.

The best reaction you can have to an obscene phone call is no reaction at all. Do not speak to the caller or answer any questions. Just hang up. If the calls continue, notify the police. Your telephone company may be able to help the police trace the calls and find and punish the person responsible for disturbing you.

Computer networks and computer bulletin boards have given some would-be child molesters a new way of finding victims. The adult meets his or her victims, usually young children or teenagers, through electronic mail. After engaging a youngster in conversations, the adult may try to arrange for a meeting.

If you are a young person who has met an adult friend on the computer, remember: a face-to-face meeting with your new friend may be dangerous. Just as you are cautious about strangers you meet on the street, you must also be cautious about strangers you meet while sitting at your computer terminal.

Think and Discuss

1. How can adults help protect children from molesters?
2. Describe a preventive strategy against date rape.
3. You work after school in a restaurant. One day, your boss says that he likes the new uniform that you are wearing. Is this sexual harassment?

SEX AND THE MEDIA

For better or worse, the media—magazines, newspapers, radio, books, television, and movies—influence many young people's ideas about sex. Don't allow the media to mold your sexual values and behaviors.

By carefully examining the words and images in magazines, on the radio, on TV, and in movies, you can become a critical consumer of the media. As you read magazine and newspaper articles, try to distinguish between what is real and what is not real. Stay alert when you watch television. Learn to distinguish real-life portrayals of sexual behaviors from script writers' fantasies.

How do companies use sex to sell products?

Suppose you are shopping for blue jeans. How do you choose which brand to buy? You probably consider comfort, cost, and appearance. Does advertising have anything to do with your decision?

Advertisements often use sex and sexual values. Unfortunately, most people do not take time to identify what it is about an ad that is appealing. They may simply go out and buy the advertised product.

Do not allow advertising to manipulate you. Try to distinguish fact from fantasy in advertising. While manufacturers can be your sources of jeans, they should not be the sources of your sexual values.

It is important to talk with someone you trust if you feel that you have been the victim of unwanted sexual attention.

What is pornography?

Pornography (pawr-NAHG-ruh-fee), or explicit sexual material, is best described by its characteristics. These include the portrayal of people, usually women, as sexual objects, rather than as sexual partners. Pornographic materials also portray sex as something that one person does to, not with, another person. Sex and violence are often linked in pornography.

People generally agree that pornography includes "dirty" books, magazines, movies, and videos. People do not agree, however, on what "dirty" means. A movie that is sexually arousing to one person may seem disgusting to another.

State and regional laws, like individuals, disagree on what is meant by pornography and what, if anything, should be done to regulate the pornography industry. The U.S. Supreme Court has ruled that local governments may use their own definitions of pornography and regulate its sale.

Think and Discuss

1. How can you prevent the media from determining your sexual values and behaviors?
2. How can sex be used to sell consumer products?
3. For your 18th birthday, an older sibling gives you a subscription to a pornographic magazine. What do you do about the subscription?

Chapter Summary

- Voyeurism and exhibitionism are unacceptable sexual behaviors that involve spying on, frightening, and hurting other people.
- Protective measures that you can take against sexual assault include loudly calling attention to exhibitionists and voyeurs, following your better judgment in reacting to uncomfortable social situations, avoiding alcohol, and learning self-defense.
- Advertising and the media influence people's ideas about sex by appealing to their fantasies.

Activities

1. Research the laws concerning sexual abuse in the state in which you live. What legal protection does a person have against an abusive spouse? What legal action can be taken against parents who abuse their children? Your local chapter of the American Bar Association is one possible source of information. A crisis center is another. Present your findings in a brief written report.

2. As a class, find out what services are available in your community to victims of rape, sexual harassment, and other kinds of abuse. Find out crisis telephone numbers, as well as the names of victims' support groups. Make a booklet of the information you gather. Have it photocopied for everyone in the class.

3. Work with three other students to design a poster that shows fallacies and facts about rape. Obtain permission to display the posters on a school bulletin board.

4. Work with one other student. Develop a role-play showing assertive ways of resisting sexual assault. Demonstrate how you would go about "saying no" in the following situations. **Situation 1:** Someone is sexually harassing you at an after-school job. **Situation 2:** An older family member is trying to persuade you to have sexual intercourse with him or her. **Situation 3:** A drunk date is trying to get you to take off your clothes. After you have completed the three role-plays, switch roles.

5. Plan a debate with your classmates on whether publication and distribution of pornographic materials should be banned. One group should support the publication of pornographic materials. The other group should support the ban on pornography.

Thinking It Over

1. Sometimes other people try to pressure you into doing something you don't want to do. It is especially hard to resist pressure if the person exerting the pressure is an employer, older relative, or other adult who has some form of authority over you. Below is a list of requests to do certain things. Copy the list of requests into your notebook. For each request, write down how you would "say no."

Requests

   a. (From an employer) Let's go to the place on the corner to get a bite to eat after work.

   b. (From an uncle) Now give me a real kiss.

   c. (From a clergy member) I only want to hold you.

   d. (From a date, in a car) Let's just lie down for awhile.

   e. (From a date, at a party) I can't hear what you're saying in all this noise. Let's go upstairs.

2. People who have been raped, or otherwise sexually abused, often feel guilty that they are somehow responsible for the assault. Unfortunately, society offers many excuses to support this fallacy. Copy the excuses below. Write one or two sentences that identify the fallacy in each excuse.

Excuses

   a. She was drunk.

   b. She was wearing sexy clothing.

   c. That child said that everything I did felt good.

   d. My parents did the same thing to me.

3. Imagine that you are a counselor at a facility where incest survivors are treated. A young man has come to you for counseling. He was abused by his father over a period of five years. Think of all the emotional needs and concerns of the incest survivor. Make a list of all these needs. How could you help the young man deal with his problems? Next to each need and concern, write a suggestion that may help him.

STD	Cause	Symptoms
Syphilis	bacteria	painless sore, rash, and fever
Gonorrhea	bacteria	**Males:** burning and painful urination; thick yellowish-green or yellowish-white discharge from penis **Females:** burning and painful urination; thick yellowish-green or yellowish-white discharge from vagina
Chlamydia	bacteria	**Males:** light, watery discharge from penis; painful urination **Females:** painful urination; discharge from cervix
HIV Infection	virus	swollen lymph glands, fever, night sweats, severe fatigue, weight loss
Herpes	virus	blisterlike sores on genitals, mouth, or face; swollen glands; fever **Males:** may have slight burning sensation when urinating **Females:** vaginal discharge
Genital Warts	virus	Warts on the penis, in the mouth, around the vagina, around the anus
Hepatitis B	virus	headaches, loss of appetite, fever, tiredness, nausea, vomiting, jaundice
Trichomoniasis	protozoa	**Males:** clear discharge, painful urination, itching **Females:** foul-smelling greenish discharge, painful itching, severe irritation around the vagina.
Candidiasis	fungus	**Males:** usually have no symptoms **Females:** thick cottage cheese-like discharge; severe itching around vagina
Pubic Lice	parasite	severe itching
Scabies	parasite	itching, discomfort, rash

Treatment	Complications
antibiotics	Untreated syphilis can cause severe illness, blindness, deafness, insanity, and eventually death.
antibiotics	Epididymitis in men and pelvic inflammatory disease (PID) in women, both of which can cause sterility.
antibiotics	Epididymitis in men and pelvic inflammatory disease (PID) in women, both of which can cause sterility.
No cure, no vaccine. Opportunistic infections treated with a variety of medications.	Eventually causes death once AIDS sets in.
Treated with the drug Acyclovir.	No cure. Virus remains in the body so reoccurrence is common.
Removed with chemicals, frozen off with liquid nitrogen, or surgically removed	Virus remains in the body. Increased chance of developing cancer of the cervix or anus.
bed rest; vaccine available	Some people suffer serious liver damage.
treated with medication	
antibiotic cream	
medicated shampoo or lotion	
hot baths, medication	

RESOURCE LIST

BOOKS

Gravelle, Karen and Susan Fischer. *Where are My Birth Parents?* New York: Walker and Company, 1993.

Parrot, Andrea. *Coping with Date Rape & Acquaintance Rape.* New York: The Rosen Publishing Group, Inc., 1988.

Salk, Lee. *Familyhood.* New York: Simon & Schuster, 1992.

The Boston Women's Health Collective, *The New Our Bodies, Ourselves.* New York: Simon & Schuster, 1992.

MAGAZINES

Straight Talk, The Learning Partnership, Pleasantville, New York.

HOTLINES

The National Child Abuse Hotline
800-4ACHILD

AGENCIES

Center for Population Options
Resource Center
1025 Vermont Avenue, NW
Suite 210
Washington, DC 20005

Incest Survivors Anonymous
P.O. Box 21817
Baltimore, MD 21222

March of Dimes, Birth Defects Foundation
Public Health Education Department
1275 Mamaroneck Avenue
White Plains, NY 10605

National Center on Child Abuse and Neglect
P.O. Box 1182
Washington, DC 20013

National Clearinghouse on Marital and Date Rape (NCOMDR)
Women's History Research, Inc.
2325 Oak Street
Berkeley, CA 94708

National Committee for Prevention of Child Abuse
332 S. Michigan Avenue, Suite 1600
Chicago, IL 60604

National Council on Alcoholism and Drug Dependence, Inc.
12 West 21st Street
New York, NY 10010
Contact them for information on alcohol- and drug-related birth defects.

National Crime Prevention Council
1700 K Street, NW, 2nd floor
Washington, DC 20006

Planned Parenthood Federation of America, Inc.
810 Seventh Avenue
New York, NY 10019
Check with your local chapter of Planned Parenthood as well.

The Alan Guttmacher Institute
120 Wall Street
New York, NY 10005

U.S. Public Health Service
Public Affairs Office
Hubert H. Humphrey Building
Room 725-H
200 Independence Avenue, SW
Washington, DC 20201

Violence Against Women Task Force
NOW Legal Defense and Education Fund
99 Hudson Street
New York, NY 10013
212-925-6635

GLOSSARY

abortion: removal of a growing embryo or fetus before it fully develops, 82

abstinence (AB-stuh-nuns): choosing not to have sex, 59

adolescence (ad-ul-ES-uns): the time between puberty and adulthood, 7

adoption: legal process by which individuals gain the right to raise someone's child as their own, 82

amniocentesis (am-nee-oh-sen-TEE-sis): prenatal test that analyzes amniotic fluid for genetic abnormalities, 46

amniotic (am-nee-AHT-ik) **sac:** fluid-filled sac that surrounds the embryo/fetus providing cushioning and protection, 42

artificial insemination (in-sem-uh-NAY-shun): process used when a couple is infertile; sperm is placed inside the woman's vagina near the cervix, 50

bisexual: a person who is sexually attracted to both men and women, 14

blastocyst (BLAS-tuh-sist): ball of cells formed by the dividing zygote, 41

blended family: a family that consists of two families combined, 75

cervical cap: firm, rubber cap that fits over the cervix; used to prevent pregnancy, 62

cervix (SUHR-viks): lower end of the uterus, 30

Cesarean (si-ZER-ee-un) **section:** the delivery of a baby through the mother's abdominal wall and uterus, 52

child abuse: physical, sexual, or emotional assault by an older person on a child, 89

chorionic villi (kawr-ee-AHN-ik VIL-i) **sampling** (CVS): prenatal test that analyzes cells from the placenta for genetic abnormalities, 46

chromosomes (KROH-muh-sohmz): structures that carry genes, 43

circumcision (sur-kum-SIZH-un): surgical procedure to remove the foreskin, 28

clitoris (KLIT-ur-us): small organ in the female that contains blood vessels and nerve endings, 31

contraception (kahn-truh-SEP-shun): the use of a number of methods to prevent pregnancy; same as birth control, 58

contraceptive sponge: dome-shaped sponge that contains a spermicide; used to prevent pregnancy, 62

corpus luteum (KOWR-pus LOO-tee-um): the follicle after fertilization, 32

date rape: forced sex between people who know each other, 93

deoxyribonucleic (dee-ahk-see-ry-boh-noo-KLAY-ik) **acid** (DNA): substance that carries the genetic code, 43

diaphragm (DI-uh-fram): dome-shaped cup that fits over the cervix; used to prevent pregnancy, 61

dilation (di-LAY-shun): first stage of labor; the opening and stretching of the cervix, 52

dilation and curettage (kyoor-uh-TAHZH) (D and C): a procedure in which the cervix is opened up and the uterus is scraped, 83

dominant gene: stronger gene in a pair, 44

ectopic (ek-TAHP-ik) **pregnancy:** pregnancy in which the zygote attaches itself to the inside of a Fallopian tube instead of the endometrium, 48

egg: female sex cell; also called an ova, 29

ejaculation (i-jak-yuh-LAY-shun): release of sperm from the penis, 26

embryo (EM-bree-oh): fertilized egg from the time of implantation to the end of the eighth week of development, 41

endocrine (EN-duh-krin) **system:** group of glands that regulate the maturation process, 5

endometrium (en-doh-MEE-tree-um): lining of the uterus, 32

epididymis (ep-uh-DID-i-mus): organ that stores sperm after they are produced and before they are ejaculated, 25

erection: state in which the penis gets larger, longer, and harder, 26

estrogen (ES-truh-jen): female hormone responsible for the growth and function of female sex organs and female secondary sex characteristics, 6:

exhibitionist (ek-suh-BISH-un-ist): person who gets sexual pleasure by showing his or her genitals, 88

103

extended family: family that includes other family members, such as grandparents, aunts, uncles, or cousins, 75

Fallopian (fuh-LOH-pee-un) **tube**: tube which connects each ovary to the uterus, 30

female condom: sheath that covers the vaginal canal, as well as the entire genital area, which prevents sperm from traveling up into the uterus; also called the vaginal pouch, 61

fertilization (fur-tul-i-ZAY-shun): when a sperm joins with an egg, 25

fetal alcohol syndrome (FAS): condition of physical, mental, and behavioral problems that affect babies born to women who drank heavily during pregnancy, 48

fetus: fertilized egg from the ninth week of development through birth, 42

foreskin: flap of tissue that covers the head of the penis, 27

gene: basic unit of genetic material carried on a chromosome, 43

gestational diabetes (jes-TAY-shun-ul di-uh-BEET-is): complication during pregnancy in which the mother's body can't produce enough insulin to control the increased level of sugar in the blood, 49

heterosexual (het-ur-uh-SEK-shoo-wul): person who is sexually attracted to members of the opposite sex, 14

homosexual (hoh-muh-SEK-shoo-wul): person who is sexually attracted to members of the same sex, 14

hymen (HI-mun): thin membrane that partially covers the vagina, 30

in-vitro (VEE-troh) **fertilization**: process used when a couple is infertile; mature eggs are fertilized in a laboratory dish with nutrients and sperm; fertilized eggs are put into the uterus, 50

incest: sexual abuse, often involving sexual intercourse, that occurs within a family, 89

infatuation (in-fach-oo-WAY-shun): short-lived, yet intense attraction toward a person whom one does not know well, 17

infertility: when a couple cannot start a pregnancy, 50

intrauterine (in-truh-YOOT-ur-in) **device** (IUD): small piece of copper or plastic inserted into a woman's uterus to prevent pregnancy, 64

labia (LAY-bee-uh): skin folds around the vagina, 30

labor: muscular contractions that soften and stretch the cervix to allow a baby to pass through, 52

male condom: sheath made of thin rubber that completely covers the penis and collects semen prior, during, and after intercourse, 59

masturbate: to stimulate your own sexual organs for pleasure, 27

menopause (MEN-uh-pawz): when a woman's menstrual cycle becomes irregular and stops, 34

menstruation (men-stroo-WAY-shun): the shedding of endometrial matter, unfertilized egg, blood, and other fluids, 33

miscarriage (mis-KAR-ij): when the embryo or fetus dies during the early part of a pregnancy and is expelled from the uterus; also called spontaneous abortion, 49

nocturnal emission: ejaculations during the night; also called wet dreams, 27

oral contraceptives: pills containing a combination of hormones which trick the body into thinking that it is pregnant in order to prevent pregnancy, 63

orgasm: muscle contractions in the genitals combined with feelings of deep pleasure, 27

ovaries (OH-vur-eez): female sex glands, 6

ovulation (oh-vyuh-LAY-shun): process by which a mature egg is released from the ovaries, 29

peer pressure: strong feeling a group puts on its members to be like the rest of the group, 9

pituitary (pi-TOO-uh-ter-ee) **gland**: gland that begins the process of physical maturation, 6

placenta (pluh-SEN-tuh): structure which allows for the exchange of nutrients, wastes, and gasses between the mother and the embryo/fetus, 41

platonic friendship: emotional bond between two people who are not sexually attracted to each other, 14

pornography (pawr-NAHG-ruh-fee): explicit sexual material, 96

preeclampsia (pree-ee-KLAMP-see-uh): complication in pregnancy characterized by rapid weight gain, swelling, and increased blood pressure, 49

premature: when a baby is born before the 36th week of pregnancy, 49

premenstrual syndrome (PMS): physical and emotional changes the body goes through before menstruation, 34

progesterone (proh-JES-tuh-rohn): hormone produced in the female's body that keeps the endometrium thick, 33

prostitution (prahs-tuh-TOO-shun): the exchange of sex for money, 88

rape: having sexual intercourse without consent, 89

recessive (ri-SES-iv) **gene**: weaker gene in a pair, 44

role model: person whom you admire and want to be like, 3

scrotum (SKROHT-um): pouch which holds the testes outside the abdominal cavity, 25

semen (SEE-mun): mixture of sperm and other liquids, 27

seminal vesicles (SEM-uh-nul VES-i-kuls): glands located under the bladder which provide most of the liquid in semen, 26

sex: being male or female; act of sexual intercourse, 2

sex role: behaviors and expectations associated with being male or female, 2

sexual harassment (huh-RAS-munt): any form of unwanted sexual attention, including leering, pinching, lewd comments, and propositions, 94

sexuality: attitudes, feelings, and values about being male or female, 2

sexually transmitted disease (STD): disease that is spread through sexual intercourse, 60

single-parent family: family in which there is only one parent present, 75

sperm: male sex cell, 25

spermicide (SPUR-muh-syd): sperm-killing chemicals, 60

stereotype (STER-ee-uh-teyp): the belief that all members of a group have the same characteristics, 4

stillbirth: when a baby is born dead, 49

testes (TES-teez): male sex glands which produce sperm and secrete hormones, 6

testosterone (tes-TAHS-tuh-rohn): male hormone responsible for development of male secondary sex characteristics, 6

tubal ligation (ly-GAY-shun): permanent method of birth control in which the Fallopian tubes are cut and tied off in order to prevent eggs from traveling into the uterus, 66

umbilical (um-BIL-ih-kul) **cord**: thick rope of blood vessels that connect the embryo and placenta, 42

urethra (yoo-REE-thruh): tube that leads out of the penis through which sperm and urine pass, 26

uterus (YOOT-ur-us): muscular, pear-shaped organ which supports and nourishes an embryo/fetus, 30

vagina (vuh-JI-nuh): passageway from the uterus to the outside of the body; sometimes called the birth canal, 30

vas deferens (DEF-uh-renz): tubes which carry sperm from the testicles to the ejaculatory ducts, 26

vasectomy (vas-EK-tuh-mee): permanent method of birth control in which the vas deferens are cut in order to prevent sperm from mixing in with the semen, 66

voyeur (vwah-YUHR): person who gets sexual pleasure from watching other people undress or have sexual intercourse, 88

withdrawal: the removal of the penis from the vagina before ejaculation, 67

zygote (ZI-gote): fertilized egg, 41

INDEX

A

Abortion, 82–83
  methods of, 83
Abstinence, birth control, 59
Adolescence
  decision making in, 8–9
  stage of, 7
Adoption, 82
Advertising, and sex, 96
Alcohol, in pregnancy, 48
Amniocentesis, 46
Amniotic sac, 42
Anesthetic, in childbirth, 51, 53
Artificial insemination, 50

B

Birth control
  abstinence, 59
  birth control pills, 63–64
  cervical cap, 62
  condoms, 59–61
  contraceptive sponge, 62
  Depo-Provera, 65
  diaphragm, 61–62
  future view, 67, 70
  intrauterine device, 64
  natural methods, 66, 67
  Norplant, 65
  pros/cons of methods, 69
  responsibility in, 58
  rhythm method, 66–67
  spermicides, 60, 62
  sympto-thermal method, 67
  tubal ligation, 66
  unreliable methods, 67
  vasectomy, 66
Birth control pills, 63–64
Bisexuals, 14, 20
Blastocyst, 41
Blended family, 75
Breastfeeding, 54

Breasts
  milk production, 55
  self–examination, 34–35

C

Cancer
  breast, 34
  testicular, 28–29
Cervical cap, 63
Cervix, 30, 53
Cesarean section, 52
Child abuse, 89–91
  emotional abuse, 90
  forms of, 89–90
  teaching children about, 90
  and victim self–blame, 91
Childbirth, 51–54
  anesthetic in, 51, 53
  Cesarean section, 54
  natural, 52
  postpartum period, 54
  process of, 52, 53–54
Chorionic villi sampling, 46
Chromosomes, 43
  and gender determination, 46–47
Circumcision, 28
Clitoris, 31
Condoms, 59–61
  female condom, 61
  male condom, 59–61
  proper use of, 59–61
Contraception, 58
  See also Birth control
Contraceptive sponge, 62
Corpus luteum, 32–33
Cystic fibrosis, 45

D

Date rape, 93
Dating, 15–17
  going steady, 16–17
  monogamous relationship, 20
Decision making, in adolescence, 8–9
Depo-Provera, 65
Diabetes, gestational, 49
Diaphragm, 61–62
Dilation, in childbirth, 53
Dilation and curettage, 83
DNA, 43
Dominant genes, 44, 45
Douching, 67
Drugs, in pregnancy, 48

E

Ectopic pregnancy, 48–49
Eggs of female, 29, 33
  fertilization of, 40–41
Ejaculation, 26–27
Embryo, development of, 41–42
Emotional abuse, 90
Emotions, in adolescence, 7–8
Endocrine system, 5
Endometrium, 32
Epididymis, 25
Erection, 26
Estrogen, 6
Exhibitionism, 88
Extended family, 75

F

Fallopian tube, 30, 33, 40, 49
Family
  blended family, 75
  extended family, 75
  importance of, 74
  single-parent family, 75
Female reproductive system, 29–30

menstrual cycle, 32–34
organs of, 29–31
Fertilization, 25
sperm and egg, 40–41
Fetal alcohol syndrome, 48
Fetus, growth of, 42
Foreskin, 27–28
Friendship, 13–14

G
Gender, prenatal determination of, 46–47
Genes, 43–44
dominant genes, 44, 45
inheritance of traits, 44
recessive genes, 44, 45
Genetic disorders
inheritance of, 45
and prenatal testing, 45–46
types of, 45
Genetics, 43–47
dominant and recessive traits, 44
gender determination, 46–47
genes, 43–44
Gestational diabetes, 49

H
Heterosexuals, 14
Homosexuals, 13–14, 20
Hormones, 5, 33
Human development
fertilization, 40–41
prenatal development, 41–42
Hymen, 30

I
Incest, 89

Infatuation, 17
versus love, 17–18
Infertility, 50–51
causes of, 50
solutions to, 50–51
Intrauterine device, 64
In-vitro fertilization, 50–51

L
Labia, 30–31
Labor, in childbirth, 53
Love
versus infatuation, 17–18
nature of, 17, 18

M
Male reproductive system, 25–29
reproductive organs, 25, 26–27
sperm, ejaculation of, 26–27
Marriage, 76–77
commitment in, 76–77
reasons for, 77
successful marriages, 77
teenage marriage, 80–81
Masturbation, 27
Media, and sex, 95–97
Menopause, 34
Menstrual cycle, 32–34
end of, 34
and fertility, 33
first period, 33
time span in, 33–34
Menstruation, 6, 33
Miscarriage, 50
Monogamous relationship, 20

N
Natural childbirth, 52

Nocturnal emission, 27
Norplant, 65
Nutrition, in pregnancy, 48

O
Obscene phone calls, 95
Obstetricians, 47–48
Oral contraceptives, 63–64
Orgasm, 27
Ovaries, 6, 29
Ovulation, 29–30, 33

P
Parenthood, 78–79
childrens' needs, 78, 79
requirement of, 78–79
Peer pressure, 9
learning to resist, 18
and sexuality, 13, 18–19
Penis, 27–28
Pituitary gland, 6
PKU, 45
Placenta, 41–42
Platonic friendship, 14
Pornography, 96–97
Postpartum period, 54
Preeclampsia, 49
Pregnancy
and childbirth, 51–54
complications of, 48–49
events of, 42–43
harmful substances in, 48
prenatal care, 47–48
Premature birth, 49
Premenstrual syndrome, 34
Prenatal care, 47–48
Prenatal tests, 45–46
amniocentesis, 46
chorionic villi sampling, 46
ultrasound, 45
Progesterone, 33

Prostitution, 88–89
Puberty, 5–8
   age of, 5
   emotional changes in, 7–8
   in females, 6
   in males, 6
   physical changes in, 5–6
   secondary sex characteristics, 6

R
Rape, 89, 91–94
   aftermath of, 94
   date rape, 93
   of men, 91
   myths/facts about, 91–92
   prevention of, 92
   statutory rape, 93
Recessive genes, 44, 45
Reproduction
   female reproduction system, 29–30
   male reproduction system, 25–29
Rhythm method, birth control, 66–67
Role model, 3

S
Scrotum, 25
Secondary sex characteristics, 6
Semen, 27
Seminal vesicles, 26–27
Sex, 2
Sex roles, development of, 2–3
Sexual abuse, 89–94
   child abuse, 89–91
   incest, 89
   rape, 91–94

Sexual deviations
   exhibitionism, 88
   prostitution, 88–89
   voyeurism, 88
Sexual harassment, 94
Sexuality
   bisexuals, 14
   heterosexuals, 14
   homosexuals, 13–14
   nature of, 2
   and peer pressure, 13, 18–19
Sexually transmitted diseases (STDs), 59, 100–101
Sickle-cell anemia, 45
Singlehood, reasons for, 77
Single-parent family, 75
Smoking, in pregnancy, 48
Sperm, 25
   fertilization of egg, 40–41
   release of, 26–27
Spermicides, 60, 62
Statutory rape, 93
Stereotype
   about sexuality, 14
   meaning of, 4–5
Stillbirth, 49
Surrogate mother, 51
Sympto-thermal method, birth control, 67

T
Tay-Sachs disease, 45
Teenage marriage, 80–81
   problems of, 80, 81
   reasons for, 80
   and teenage pregnancy, 80–81
Testes, 6

Testicles, self-examination, 28–29
Testosterone, 6
Tubal ligation, 66

U
Ultrasound, 45
Umbilical cord, 42
Urethra, 26, 27
Uterus, 30, 33

V
Vacuum aspiration, 83
Vagina, 30–31
Vas deferens, 26
Vasectomy, 66
Voyeurism, 88

W
Withdrawal, as birth control, 67

Z
Zygote, 41